Zoltan J Kiss

Quantum Energy and Mass Balance

The Gift of the Earth

2009

Order this book online at www.trafford.com/08-1547
or email orders@trafford.com

Most Trafford titles are also available at major online book retailers.

Note for Librarians: A cataloguing record for this book is available from Library
and Archives Canada at www.collectionscanada.ca/amicus/index-e.html

ISBN: 978-1-4251-9157-3

*We at Trafford believe that it is the responsibility of us all, as both individuals
and corporations, to make choices that are environmentally and socially sound.
You, in turn, are supporting this responsible conduct each time you purchase a
Trafford book, or make use of our publishing services. To find out how you are
helping, please visit www.trafford.com/responsiblepublishing.html*

*Our mission is to efficiently provide the world's finest, most comprehensive
book publishing service, enabling every author to experience success.
To find out how to publish your book, your way, and have it available
worldwide, visit us online at www.trafford.com/10510*

www.trafford.com

North America & international
toll-free: 1 888 232 4444 (USA & Canada)
phone: 250 383 6864 ♦ fax: 250 383 6804
email: info@trafford.com

The United Kingdom & Europe
phone: +44 (0)1865 487 395 ♦ local rate: 0845 230 9601
facsimile: +44 (0)1865 481 507 ♦ email: info.uk@trafford.com

10 9 8 7 6 5 4 3 2 1

for Anna and Jozsef

Executive Summary

The *quantum energy* impact of the *Earth* is available for us to use.

Matter exists in two forms: mass and energy. In this book we examine the balance of their permanent interaction.

We continue the discussion about the energy balance of relativity, the subject of the first book published in 2007*. This is why the book starts with section 13. A summary of the first 12 sections is given in the attachment.

The quantum energy and mass balance is the basis of the structure of the elementary world. Elements are the results of the transformation of mass into energy and the retransformation of energy into mass.

The key is the sphere symmetrical expanding acceleration for infinite time, the motion with $i = \lim a\Delta t = c$.

The Pound-Rebka-Snider experiment at Harvard University provides evidence of the *gravitational* shift of frequencies of the radiation to and from the surface of the *Earth*. This experiment is concerned with the measured single shift, either *blue* or *red*. In order to make it in the correct way we have to measure it in full: the *red* shift of the *blue* shifted reflection.

The explanation is simple. As an effect of *gravitation* there is no difference in the shifts of the reflected and original frequencies. The blue and red shifts are equal. As a result of the sphere symmetrical expanding *acceleration* of the *Earth,* however, the red shift of the reflected frequency should be more then the blue shift. This gives working capability.

The measured energy intensity *surplus*, as the first book predicted, is the benefit of the motion of the *Earth* available for us to use.

* *The Energy Balance of Relativity, the Theory of Event Concentration and Acceleration for Infinite Time*
Trafford Publishing 2007; ISBN 1-4251-1502-0
www.trafford.com/06-3261

Everything starts with the definition of *time*.

What is *time*? Can time be defined without an event? What is the *all-over-acting-universal-event* which establishes time?

Trying to answer these questions we come to the conclusion that, without an event, *time* as such cannot be identified. The only natural and universal *event* "establishing time" is the transformation of the mass into energy and the retransformation of the energy into mass.

No event would mean *no time*. And *no time* would mean *no matter*.

We think we easily detect the mass part of matter. We must not forget, however, that we measure, in fact, the *effect* of the mass rather than the mass itself. But where is the energy part?

Once the mass has been transformed into energy quantum *the energy* is represented by photons of the *Quantum System of Reference*.

The general rule of *entropy* prevents the full transformation of mass into energy and the full transformation of energy into mass. The *quantum entropy* is the smallest portion of the matter in mass status.

The adiabatic box of matter with infinite boundaries is in mass and energy balance. *Mass*, in sphere symmetrical expanding acceleration and accelerating collapse, collides with photons.

Can the collision of mass with photons result in the increase of the energy of photons, all of equal energy quantum? Obviously not.

The motion of photons with *c*, the speed of light is the status of the *Quantum System of Reference*. This speed value, however, means much more than "just" being the speed of light.

The collision of mass with photons is energy exchange. But photons are of constant *c* speed and of constant and equal energy quantum. How can they participate in energy exchange?

c, the speed of photons, stays constant – *but the number of photons in collision* – the frequency – *will be changing*. The motion of mass, as a consequence of the collision either slows down or speeds up. The frequency corresponds to the time flow and the time flow is a consequence of the motion of the mass. As a result, there will be either more or less collision for the unit period of time. Cause and consequence exist together.

Are photons – impacted in collision with mass – "travelling" in certain directions with a mission?

Are they carrying the mass impact of the collision? Not at all. It would obviously mean an "energy content" difference between photons.

Photons are of equal energy quantum. They do not carry the mass impact. Instead they collide with each other and transfer the impact in any direction with c, the speed of photons. This is why this "impact-signal transfer" has no vector components. Photons transfer any impact (not just the light) with c, the speed of light.

The impact to the photons is transferred as by a membrane: the *Quantum Membrane*.

The *Quantum Membrane* transfers impacts, but how far? The distance is proportional to the intensity (energy) of the impact. It will either be received or it will be dissipated within the *Quantum System of Reference*.

In our conventional understanding, energy is something that has been spent or lost, which disappears, giving in return results, for example, motion or work. The existence of the matter in time, its quantum energy and mass balance proves that *energy* cannot be lost.

Mass used for its own acceleration will result in having less and less value. It leads to an ever fewer number of collisions with photons of the *Quantum System of Reference*. It finally reaches the mass value of quantum entropy, the last mass status of the matter. But the energy must also be present within the infinitely large adiabatic box of the matter. A part of matter during the mass transformation cannot disappear!

The energy is not just present! During the transformation of mass into energy there will be a point when the impact of photons of the *Quantum System of Reference*, in permanent interaction with mass, prevails and the process turns around: Under the impact of the motion of photons of the *Quantum System of Reference*, the quantum entropy collapses and the matter transforms from its energy status into mass.

The retransformation of energy into mass is the sphere symmetrical accelerating collapse. There is an entropy status at this end as well. There is no pure energy without mass and there is no pure mass without energy.

Why are the sphere symmetrical expanding acceleration and accelerating collapse the only forms of the existence of the mass status of the matter?

The motion of mass with constant speed is relative. It is identical to a stationary status or to a status of rest. Both, the relative and the absolute statuses of rest would mean no impact, "no event". Therefore, the acceleration is obligatory. Nothing else can give the natural freedom to the motion, just the sphere symmetrical.

There are an infinite number of mass particles in transformation and re-transformation. The *Quantum System of Reference* is common. There should be a structure in this all over mass-energy balance.

And we do indeed find elements, with *protons, neutrons, electron* and other particles.

The *proton* is the transformation of the matter from its mass status into its energy status. The proton is the event of the sphere symmetrical expanding acceleration of the mass up to the speed $i = \lim a\Delta t = c$.

How can the *proton* be an event?

The explanation here needs the acceptance of the existence of an infinite number of different time systems.

Mass particles of atoms of elements are in sphere symmetrical expanding acceleration (and collapse) in motion with $i = \lim a\Delta t = c$ relative to the *Earth*. The duration of the events with speed $i = \lim a\Delta t = c$ within the elements is infinitely long relative to the measured duration of the same events in the system of reference on the surface of the *Earth*. Events of the elementary world are infinitely short measured effects within the system of reference of the *Earth*. And we have to remember: the *measured* mass (the weight of a particle) represents the effect of the mass, which is an event.

The *electron* is the measured effect of the sphere symmetrical expanding acceleration of the mass, in motion with $i = \lim a\Delta t = c$. The electron is in collision with photons. The process is balanced by the *Quantum Membrane* until the mass reaches the status of quantum entropy. The collision of mass with photons in the electron process results in blue shift.

The turning point is the status of quantum entropy, the smallest existing mass of the matter. The mass is collapsing under the "pressure" of the impact of the photons (energy) of the *Quantum System of Reference*. The sphere symmetrical collapse of the quantum entropy is the *neutron* process.

The mass-energy balance of the proton-electron-neutron process is the cohesive force of atoms. Damages to this balance produce certain impacts to the *Quantum Membrane*. (In conventional meaning these are particles and radiation.) Mass unbalance in the cycle results in *alpha* radiation. Deviation in the proton process results in *beta*, deviation in the neutron process results in *gamma* radiation.

What about the timeframe of the starting status of the *proton* and the final stage of the *neutron* processes? These are close to the time count of the system of reference of the *Earth*. Here the explanation is easy: this is the reason we find (the effect of) these particles.

The cycles together compose all elements with identified and measured proton mass effects (weight values), as presented by the Periodic Table.

The last element of the chain is *Hydrogen*. *Hydrogen* has *no* measured *neutron* mass effect. It means in conventional terms that *Hydrogen* does not have a neutron. The point is that the duration of its neutron process is infinitely long with infinite low intensity. The infinite low intensity results in fact in "no effect" and no (measured by us) *Hydrogen* neutron mass.

Should the neutron process of the *Hydrogen* have an end it would mean that the matter would also have its end. It would be the final step of the cycle of the transformation of the matter. The cycle cannot end with *zero* neutron mass. It would contradict the general rule of entropy. There would be no room for continuation. Matter cannot have its end.

The proton and neutron processes must be in balance. They represent the opposite parts of the cycle: expanding acceleration versus accelerating collapse. They are balanced, but with different intensities (measured effects) of the process.

The electron blue shift is the driving force of the neutron collapse. If the neutron process is less intensive than the proton process, the generating blue shift within the element can be used by neutron red shifts of other elements. The blue shift is indicating the level of the activity of the element in reactions with others. The intensity of the blue shift is the indicator of the energy source of the element.

Elements with blue shift surplus are energy providers. The most important of them are *Hydrogen* and *Oxygen*. All life important elements, however, have blue shift surplus.

The intensity of the blue shift of the *Hydrogen* is extremely high. This means that the intensity of its electron process is extremely high and results in extremely high effect – measured electron weight.

Matter exists independently of us people. We have to distinguish genuine (natural) events that happen on account of the *internal* energy of the mass (of the matter) from those that happen under the effect of the *external* energy (work) of people.

Both result in motion. In the first case the mass transforms into energy and becomes less. In the second the mass does not change – but its effect does. It is measured as work, spent within the system of reference of the *Earth*.

Elements are stable in their established proton-electron-neutron process relations. Molecules of compounds with different elements are formulating on the basis of their common blue shift balance.

Elements with blue shift surplus can provide their available blue shift to those others which are capable of using it.

Atoms of elements with blue shift surplus are in blue shift conflict with each other. This is the reason we find elements in gaseous phase in normal temperature circumstances.

The blue shift conflict keeps atoms in motion. Solid, liquid and gaseous statuses are determined by blue shift thrust between atoms of elements and molecules of compounds. The *plasma* phase is the motion of atoms in blue shift conflict with speed of $\lim i = c$.

H_2O molecules are based on the blue shift of the *Hydrogen* and are liquid at normal room temperature. The blue shift of the *Hydrogen* intensifies the neutron process of the *Oxygen*. The blue shift of the *Oxygen* also affects, but the *Hydrogen* neutron collapse remains infinitely low as it was.

In *Hydrocarbon* molecules the *Hydrogen* intensifies the blue shift of the *Carbon* atom, which results in blue shift conflict between the molecules of the *Hydrocarbon* and brings it to liquid status.

The aggressive blue shift surplus of the *Oxygen* destroys elements and compounds. The destruction of the elementary structure means that the blue shift process happens within the system of reference of the *Earth*, with time flow infinitely slow and with intensity infinitely high relative to the elementary system of reference. The consequence is *Fire*, electron blue shift on the *Earth* of infinite frequency.

The destruction of the elementary structure is the basis of *Hydrocarbon* and *Nuclear* "quantum engines". The *Fire* is destruction and the nuclear chain reaction is destruction as well. Electric engines are quantum engines: wheels are rotating as a result of the blue shift conflict of electrons.

Elements represent the cycles of the transformation of the matter. But we only measure integer values of the proton mass effect (weight). What is between?

The effect of the proton transformation within the cycle with more mass but with less intensity is identical to the effect of less mass with higher intensity. It means we can easily measure identical integer effects (weight values) with – in fact – changing proton masses.

More proton mass always means electron dominance and blue shift surplus. Lesser proton mass results in electron shortage and always means blue shift deficit. These are *magnets*. There is blue shift dominance at one end and blue shift deficit at the other. The intensity of the blue shift varies along the length of the *magnets*.

The sequence of the proton mass change in magnets happens in space instead of time. Magnets resolve their unbalance by their *magnetic blue shift field* through the *Quantum System of Reference*.

*E*arth (part of the matter) is in sphere symmetrical expanding acceleration. The acceleration is *g*, the speed is constant, value of $i = \lim g\Delta t = c$. The constant speed during the acceleration is the result of the balance with the impact of the collision with photons of the *Quantum System of Reference*.

Let us impact the *Quantum Membrane* at a certain level above the *Earth*. The *Quantum Membrane* transfers the impact to the *Earth*. The impact will be blue shifted by the collision with the surface of the *Earth*. The frequency of the collision will increase.

The impact, reflected from the surface of the *Earth*, will be red shifted by the detection, at the level of the original impact. The frequency will be less than the frequency of the original impact.

The value of the red shift at the detector is more than the value of the blue shift at the reflection. This gives energy intensity surplus.

The sphere symmetrical expanding acceleration of the *Earth* is a gift. This gift is an energy intensity benefit, the result of the *quantum energy* impact of the *Earth* – a gift to be used.

Table of Contents

13

Transformation of mass into energy

$\mathbf{W}$e examine the mass-energy balance of the motion when external and internal energies are used.

13.1
The mass balance of motion when *external energy* is used

We take a stationary system of reference with time measurement t_o and denote it as *SORto*. We also take a system of reference in acceleration within *SORto* and mark it as *SORtv*. We take *SORtv* in three statuses: at rest and denote it as *SORtvo*, and in motion at speed v_n and v_k denoting it through *SORtvn* and *SORtvk*. We assess the impact of the motion on the value of a *mass*, placed within the system of reference. The examination is made at particular time moments when the actual speed values are different but taken as constant for the examination.

The value of the mass is m and it is stationary relative to the system of reference in motion. The energy source, which accelerates the system of reference, is unlimited and external relative to the mass, placed within the system of reference.

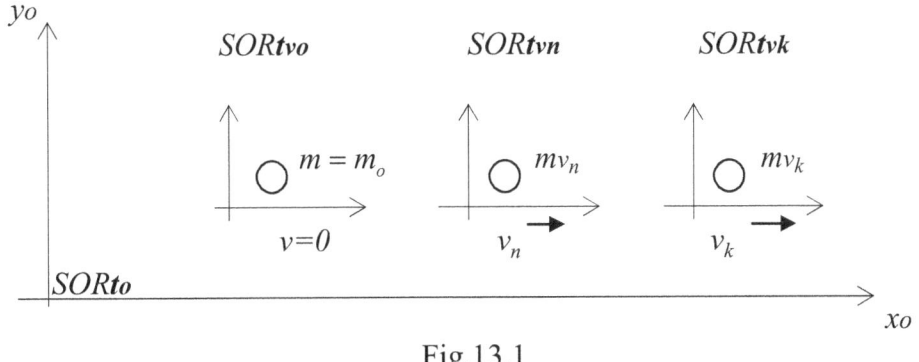

Fig.13.1

Fig.
13.1

Fig.13.1 shows *SORto*, the stationary system of reference, containing the other three statuses of the system of reference in motion:

- *SORtvn*, with speed v_n at time moment t_{vn};
- *SORtvk*, with speed v_k at time moment t_{vk};
- *SORtvo* is the stationary status with *v=0* at rest within *SORto*.

We also take that $v_k > v_n$ and $v_n = at_{vn}$; $v_k = at_{vk}$
a – is the acceleration,

We are looking for the mass value within the system of reference at $v = 0$ and in motion at speed v_n and v_k. We denote the value of mass m, stationary within *SORto* through m_o.

(We have to remember here, that there is no device which could measure the mass directly. We usually measure the effect of the mass in certain circumstances and make an assumption on its value upon the results.)

Are the mass values different *within SORtvo* at rest and *within SORtvn* and *SORtvk* in motion with v_n and v_k? The answer is unambiguously *no*. They are *the same*, equal to each other and to the mass value of the stationary status:

13A1

$$m_{vn} = m\sqrt{1-\left(u^2/c^2\right)}; \text{ and } m_{vk} = m\sqrt{1-\left(u^2/c^2\right)}$$

13A2

$$m = m_{vn} = m_{vk} = m_o \qquad\qquad \text{and } m = m_o\sqrt{1-\left(u^2/c^2\right)}$$

where *u=0* means, the mass placed within *SORtv* is stationary during the motion (in positions *SORtvn* and *SORtvk*).

The answer comes from the reciprocal character of the relative motion.

The constant speed of the motion at any particular time moment of the examination between *SORtvo*, the stationary status and *SORtvn* and *SORtvk* the statuses of the system of reference in motion, are reciprocal. The value of the mass within the systems of reference in motion must also satisfy the case when the systems of reference in motion are supposed to be the stationary ones and *SORtvo* is supposed to be the one in relative motion. There is no reason to "measure" different mass values within a stationary system of reference while other systems are in motion

(This principal relativistic approach does not change however the fact: to accelerate the systems of reference – *SORtvn* and *SORtvk* – to higher speed requires more *work*. The effects of the same mass are different.)

Ref
5H1

With reference to 5H1, the work to be envisaged in *SORto*, for accelerating *SORtvn* up to v_n and *SORtvk* up to v_k is:

13A3

$$W_o = 0; \qquad W_n = m_{vn}c^2\left(1-\sqrt{1-\frac{v_n^2}{c^2}}\right) \qquad \text{and} \qquad W_k = m_{vk}c^2\left(1-\sqrt{1-\frac{v_k^2}{c^2}}\right)$$

where $m = m_{vn} = m_{vk}$ is the mass value within the statuses of the system of reference in motion. While the mass values are equal, the work values, since $v_k > v_n$, are obviously $\quad W_k > W_n$

The *work intensities* in 13A1, with reference to Section 5.03, are:
- from the point of view of *SORtvn* and *SORtvk*, the internal systems of reference:

$$w_{vn} = m_{vn}c^2\left(1 - \sqrt{1 - \frac{v_n^2}{c^2}}\right) \quad \text{and} \quad w_{vk} = m_{vk}c^2\left(1 - \sqrt{1 - \frac{v_k^2}{c^2}}\right) \quad \begin{array}{l} \text{13A4} \\ \text{13A5} \end{array}$$

- from the point of view of *SORto*, the external system of reference:

$$w_{on} = \frac{m_{vn}c^2}{\sqrt{1 - \frac{v_n^2}{c^2}}} - m_{vn}c^2 \quad \text{and} \quad w_{ok} = \frac{m_{vk}c^2}{\sqrt{1 - \frac{v_k^2}{c^2}}} - m_{vk}c^2 \quad \begin{array}{l} \text{13A6} \\ \text{13A7} \end{array}$$

As a conclusion we can state that in the case of motion, caused by external energy (different than that of the internal energy of the mass), equal mass values within the stationary system of reference will also be equal and the same within systems of reference in motion.

> The effect of the same mass however results in different absolute work values and work intensities.

13.2

The mass balance, when the *internal energy* of the mass is used

S.
13.2

We are taking in this case a mass, value of m_o at rest, and placing it within a stationary system of reference. The mass has an infinite number of mass components inside its structure. In order to distinguish different mass values and mass components, we will call the full mass, placed within the stationary system of reference, *matter*.

We suppose that at this stationary status, all mass components inside the matter are also stationary. We may not be aware of the internal structure of the mass composition, but the sum of all mass components inside the matter must give m_o the full value of the mass at rest:

$$m_o = m_1 + m_2 + ... + m_{n-1} + m_n \quad \text{and} \quad \text{13B1}$$

$$m_1 = m_{1_1} + m_{12} + ... + m_{1_{n-1}} + m_{1n} \quad \text{and} \quad \text{13B2}$$

....

$$m_{n_n} = m_{n n_1} + m_{n_{n_2}} + ... + m_{n_{n_{n-1}}} + m_{n_{n_n}} \quad \text{and so on} \quad \text{13B3}$$

With reference to Section 5.1.1, the absolute energy of the matter is:

$$E_o = m_o c^2 \quad \text{13B4}$$

What is the mass value of the matter if the mass components within the matter are in motion?

For the better understanding of the case, we imagine the stationary system of reference, where the matter is placed, as a virtual box with no mass and with unlimited borders. We are looking for the value of the mass within this virtual stationary box-system-of-reference, when all mass components are in motion.

S.
13.2.1   *13.2.1. The measured mass of the matter in motion*

What could be the motion generated by the internal energy of the matter? We examine the sphere symmetrical expanding acceleration of each mass component. This is in fact the expansion of the mass particles of the matter relative to their own mass centre. A schematic outline is shown in Fig.13.2.

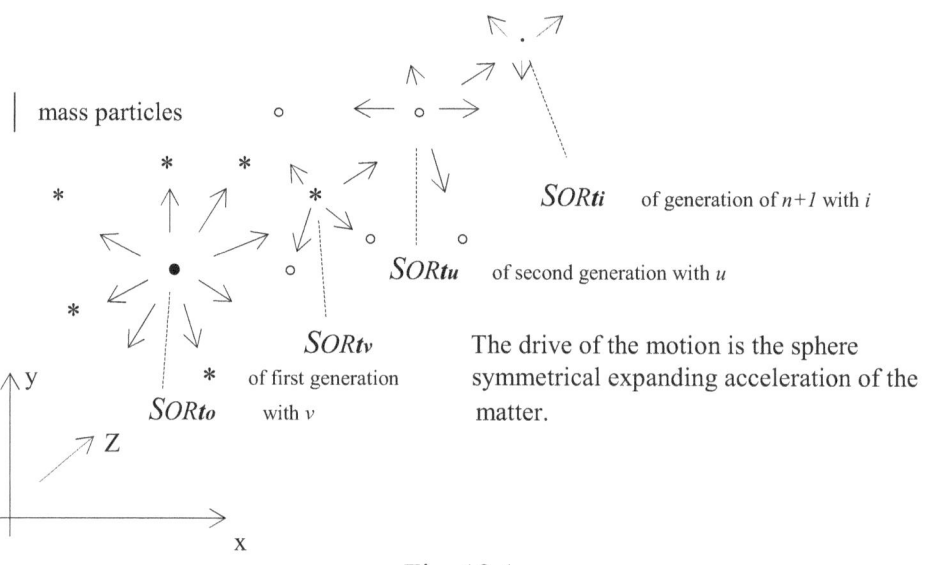

Fig
13.1

Fig. 13.1

Once the expanding acceleration starts, there will be an infinite number of mass components and particles in motion. Each and all of them can be considered as an origin of further mass expansion, with generation of new mass particles. Since the energy of mass is transforming into energy of motion, the mass of these particles in sphere symmetrical expanding acceleration is expected to be always less than the mass of their origin.

There could be an infinite number of levels during the transformation, showing the sequence of the generation of sub-, and sub-sub-particles through the sphere symmetrical expanding acceleration of the matter. Each level could represent a certain status of the transformation, where the summarised mass value generating at each consecutive level was less and less until the energy of the mass is sufficient for transformation.

Fig 13.2 is a simplified projection of the sphere symmetrical expanding acceleration of the matter from its stationary *SORto* status, through an infinite number of intermediate stages, finally into motion with $i = \lim a\Delta t = c$, *SORti*. Systems of reference *SORtv* and *SORtu* represent the acceleration of the mass components at actual speed v and u. Together with *SORti* they mark different levels of the motion of mass particles.

For simplicity of comparison, we denote the stationary status of the matter, mass value of m_O, by *SORto*, and call it stationary system of reference. The expanding acceleration means particles with mass, less than their original expanding mass value, speeding away from the centre of the acceleration in any direction. Each particle, the result of the acceleration, becomes a new centre of sphere symmetrical acceleration.

The levels of the transformation are denoted through various systems of reference. *SORtv* marks all those mass components or particles with speed v which are the result of the sphere symmetrical expanding acceleration of the mass of the stationary system of reference. *SORtu*, means the next level, the system of those particles, which are "born" as the result of the further expanding acceleration of the matter from v to u. The last level is marked as *SORti*, where the speed is $i = \lim a\Delta t = c$.

This explanation seems to be very hypothetical, but there is no other way to describe the process. Why?
- Because with reference to Section 1.3.1 and formula 1X8, without an event (motion) the time is indefinable. (No event means no time.) Therefore, all mass components must be in motion;
- Because any other possibility of the transformation can either be excluded or covered by the sphere symmetrical expanding acceleration:

  ➤ The motion with $\dfrac{dv}{dt} = 0$ is not excluded, but this is not about   13B5

    energy transfer, either internal or external. It is a status of $\dfrac{dw}{dm} = 0$,   13B6

    the constant energy. We will prove later that while theoretically it cannot be excluded, this can hardly occur until the full energy of the mass has been exhausted.

  ➤ Until $\dfrac{dv}{dt} > 0$ is valid, the direction of the acceleration of the   13B7

    particles in fact is irrelevant. The sphere symmetrical type represents any direction perpendicular to the supposed spherical surface of the mass components in transformation. Therefore, the

    correct expression of 13B7 is $\dfrac{d|v|}{dt} > 0$. This excludes the case with   13B8

13B9   ➤ $\dfrac{dv}{dt} < 0$. The sphere symmetrical collapse is a different category of the energy transformation.

          ➤ Collision of mass particles with each other is an important factor, but in this case it is irrelevant, and does not influence in substance the results of the transformation.

          ➤ We could also imagine a "rocket type" acceleration of the mass components of matter, which might accelerate mass particles in a variety of random directions. The need to use the internal energy of the mass, the transformation of the mass into motion, for making the acceleration to happen, would have, however, been the same. The "rocket type" acceleration of the mass components of any mass particles in any direction is in fact identical to a sphere symmetrical expanding acceleration.

Since only the internal energy of matter is available, should any motion occur, only this internal energy can be its source. The matter is *adiabatic*, with no energy exchange with the outside environment - should it be that the outside environment as such, out of the unlimited boundaries of the stationary system of reference, even exists at all.

Ref 13B4   The energy of the matter within the stationary system of reference, with reference to 13B4, is:

$$E_o = m_o c^2$$

Ref S.6 13B4   With reference to Section 6, in systems of reference we measure energy intensities, function of the time flow, therefore, 13B4 can be written as:

13C1
$$e_o = m_o c^2 - m_o c^2 \sqrt{1 - \frac{v^2}{c^2}} + m_o c^2 \sqrt{1 - \frac{v^2}{c^2}}$$

The definition of the components of 13C1 will be given later in this section.

For the matter with mass components at rest, *v=0*, if there is no motion at all, 13C1 gives 13B4, the "measured" mass: m_o

Once the system of reference is in motion, m_o the mass of the stationary system of reference is being transformed into motion, creating with that *SORtv*, the system of reference in motion with *v*, replacing the original *SORto*. The description of the energy of *SORtv*, the system of reference in motion, the result of the sphere symmetrical expanding acceleration of the matter from its stationary status up to speed *v*, is:

13C2
$$e_v = m_v c^2 - m_v c^2 \sqrt{1 - \frac{v^2}{c^2}} + m_v c^2 \sqrt{1 - \frac{v^2}{c^2}}$$

Ref S.6.4   With reference to Section 6.4, the first two components of the right-hand side of the equation in 13C2 characterize the kinetic energy of the motion, the result of the accelerating work of the internal energy of the matter:

$$\Delta e_{v-kin} = w_v = m_v c^2 - m_v c^2 \sqrt{1 - \frac{v^2}{c^2}}$$ 13C3

The third component of the right-hand side gives the value of the remaining energy reserve at rest of the system of reference in motion:

$$e_{v-rest} = m_v c^2 \sqrt{1 - \frac{v^2}{c^2}}$$ 13C4

13C3 and 13C4 together give the full energy of the matter in motion:

$$e_v = \Delta e_{v-kin} + e_{v-rest}$$

Since the matter is taken as adiabatic, the full energies of the stationary system of reference and the system of reference in motion must be equal.

Ref
S.6

With reference to Section 6 on the values of the event concentration, the equations in 13C1 and 13C2 are only comparable if they are adjusted by the values of the event concentration of the systems of reference. The relation of the event concentration of systems of reference is the inverse of the time relations. If we take for the value of the event concentration of the stationary system of reference: $z_o = 1$,

the event concentration of the system of reference in motion with v relative to the stationary one is: $z_v = \sqrt{1 - v^2/c^2}$

The energy balance between the two statuses is

$$\frac{m_o c^2}{z_o}\left(1 - \sqrt{1 - \frac{v^2}{c^2}} + \sqrt{1 - \frac{v^2}{c^2}}\right) = \frac{m_v c^2}{z_v}\left(1 - \sqrt{1 - \frac{v^2}{c^2}} + \sqrt{1 - \frac{v^2}{c^2}}\right)$$ 13C5

Since for the stationary system of reference the event concentration is taken as $z_o = 1$, the absolute energy and work values are equal to their intensities.

Ref
S.5

With reference to Section 5 and 5D4, $m_v = m_o\sqrt{1 - v^2/c^2}$

5D4

13C5 proves the identical absolute energy balance of the two statuses.

The mass within *SORtv*, with reference to 13C2, is: m_v 13C6

With reference to 5D4 as above, the mass in 13C6 is equal to

$$m_v = m_o\sqrt{1 - v^2/c^2}$$ 13C7

which at *v=0* is equivalent as expected to m_o

Since the energy resource of the motion is the internal energy of the matter, the value of the mass in motion with reference to 13C7, is less than in its stationary status. The missing part is transformed into motion.

S. 13.2

13.2. *The value of the mass, transformed into motion*

The mass, transformed into motion, is obviously the difference between the mass values of the two statuses (or systems of reference) at rest and in motion of the matter:

13D1
$$m_{tr} = m_o - m_v \, ; \quad \text{which is equivalent to} \quad m_{tr} = m_o - m_o\sqrt{1 - \frac{v^2}{c^2}}$$

where v is the speed of *SORtv* relative to *SORto* (which actually does not exist anymore)

Thus, the internal energy that transforms the matter from the status at rest into motion with speed $v = at$ is:

13D2
$$\Delta e_{tr} = m_o c^2 - m_o c^2 \sqrt{1 - (v^2/c^2)} \quad \text{or} \quad \Delta e_{tr} = m_{tr} c^2$$

13D1 shows that this is the kinetic energy of *SORtv*, the matter in motion, just expressed through the parameters of the stationary system of reference:

13D3
$$\Delta e_{o-kin} = w_o = m_o c^2 - m_o c^2 \sqrt{1 - \frac{v^2}{c^2}}$$

In this case the remaining energy reserve at rest of *SORtv*, expressed through the parameters of the stationary system of reference, with reference to 13C7, does indeed give the remaining "stationary" mass value of *SORtv*:

$$m_v = m_o \sqrt{1 - v^2/c^2}$$

There is no contradiction between 13C3 and 13D3, merely the formulas refer to different systems with different values of event concentration. With reference to 13C5, the absolute energy and work values are equal.

With reference to 13A4 and 13A5, the kinetic energy and the work intensity formula of the motion, result of the use of external energy, would be

$$\Delta e_{o-kin(external)} = w_{o(external)} = \frac{m_v c^2}{\sqrt{1 - v^2/c^2}} - m_v c^2$$

with the clear indication of the mass at rest equal to m_v

S. 13.2.3 Ref S.6

13.2.3. *The balance of the transformation*

With reference to Section 6, we measure the appearance of the events, energy and work intensities, rather than their absolute values in systems of reference. Events happen not just for different time periods, but with different intensities in systems of reference distinct in motion.

The sphere symmetrical expanding acceleration of matter happens simultaneously in an *infinite* number of systems of reference, all different in their statuses and speed values of motion - at different stages of the sphere symmetrical expanding acceleration. While the event is the same, the time and energy relations of these systems of reference are different.

The energy intensity and the work intensity of the acceleration within, or relative to
- *SORto*, the stationary system of reference are:

$$e_o = \frac{dm_o}{dt_o} c^2 \; ; \quad w_o = \frac{dm_o}{dt_o} c^2 \left(1 - \sqrt{1 - \frac{v^2}{c^2}} \right)$$ 13E1

- *SORtv*, the system of reference in motion with $v = at$

$$e_v = \frac{dm_v}{dt_v} c^2 \quad w_v = \frac{dm_v}{dt_v} c^2 \left(1 - \sqrt{1 - \frac{v^2}{c^2}} \right)$$ 13E2

where $dt_v = \dfrac{dt_o}{\sqrt{1 - v^2/c^2}}$

What is the difference in energies between *SORto* and *SORtv*, the two statuses of the matter in transformation?

We cannot compare the energy values in 13E1 and 13E2 directly, since events in the two systems of reference happen in different time *horizons*. The comparison of events simultaneously happening in *SORto*, *SORtv* (and other systems of reference) describing different stages and levels of the sphere symmetrical expanding acceleration of the matter, with reference to 13C5, is only possible if work and energy intensities of the different systems of reference are adjusted by *z,* the value of the *event concentration.*

For simplicity, we take the event concentration of the stationary system of reference is equal to $z_o = 1$.

Ref
S.6

With reference to Section 6, the event concentration is the inverse of the time relations. Its value for *SORtv*, the system of reference representing the stage of the acceleration of the matter with mass components in motion with *v*, will be: $\quad z_v = \sqrt{1 - v^2/c^2}$

The correction by the corresponding values of the event concentration brings the values of different time flows to common, absolute or a supposed-to-be absolute dimension:

$$\Delta E = \frac{m_v c^2}{z_v} - \frac{m_o c^2}{z_o} = \frac{m_o c^2 \sqrt{1 - \left(v^2/c^2\right)}}{\sqrt{1 - \left(v^2/c^2\right)}} - \frac{m_o c^2}{1} = 0$$ 13E4

Through the values of the event concentration, the energy intensities in 13E4 are transformed into absolute energy values, comparable with each other.

The zero energy difference between the stationary system of reference and the one in motion proves: the energy has not been lost, just transformed. In spite of the difference in the statuses of the matter and in the time flows, the value of the absolute energy, corrected by event concentration, is equal:

$$\frac{dm_o c^2}{z_0} = \frac{dm_v c^2}{z_v} \quad \text{or} \quad \frac{m_o c^2}{z_0} = \frac{m_v c^2}{z_v}$$ 13E5

Substituting $m_v = m_o\sqrt{1 - v^2/c^2}$ and $z_v = \sqrt{1 - v^2/c^2}$ 13E5 gives equivalence.

We can write 13E5 in a different form, adding to and subtracting from the right-hand side of the equation the remaining energy reserve of rest of mass m_v in motion:

13E6
$$\frac{m_o c^2}{z_0} = \frac{m_v c^2}{z_v} - \frac{m_v c^2}{z_v}\sqrt{1 - \frac{v^2}{c_2}} + \frac{m_v c^2}{z_v}\sqrt{1 - \frac{v^2}{c^2}}$$

In 13E6 the absolute value of the kinetic energy of the mass converted into motion, with reference to 13C3 is: $E_{v-kin} = \frac{m_v c^2}{z_v} - \frac{m_v c^2}{z_v}\sqrt{1 - \frac{v^2}{c_2}}$

and its remaining energy reserve at rest: $E_{v-rest} = \frac{m_v c^2}{z_v}\sqrt{1 - \frac{v^2}{c^2}}$

We can write 13E6 also in form:

13E7
$$\frac{m_o c^2}{z_o} - \frac{m_v c^2}{z_v}\sqrt{1 - \frac{v^2}{c^2}} = \frac{m_v c^2}{z_v} - \frac{m_v c^2}{z_v}\sqrt{1 - \frac{v^2}{c_2}}$$

As for the simplicity of the equations above, the differential mark is left out. We shall not forget, however, that this is a continuous process where the dm and dm_v would be the only way to present the case. We will return to such a description from time to time.

13E7 gives a simple function of the transformation, with the meaning:

The kinetic energy of the matter in motion with v, or the utilized work of the transformation of the matter for getting the stationary mass m_o into motion with speed v (the right side of 13E7) is equal to the difference between the total energies of the masses of the matter within its two statuses, the two systems of reference in relative motion (left hand side of 13E7):

13E8
$$m_o c^2 - m_v c^2 = \frac{m_v c^2}{z_v} - \frac{m_v c^2}{z_v}\sqrt{1 - \frac{v^2}{c_2}}$$

We have to note here that the left-hand side part of 13E7 and 13E8 connects two systems of reference with different time flows, which means: *the transformation needs time, and happens for a measurable time period.* In other words: The transformation of the stationary mass into motion is the prerequisite of the time flow.

We can continue with the transformation process, involving the next levels of the transformation, the next systems of reference of the matter. Should mv of *SORtv* be transformed into motion with u within the frame of *SORtu*, (transformation of mv in motion with v into mu in motion with u) the equation of this process would be:

13F1
$$\frac{dm_v c^2}{z_{vo}} = \frac{dm_u c^2}{z_{vu}} \qquad \text{or} \qquad \frac{m_v c^2}{z_{vo}} = \frac{m_u c^2}{z_{vu}}$$

where m_u is the mass of the system of reference, in motion with speed u.

The event concentration of *SORtv*, the system of reference in motion at v, can be taken in this case as $z_{vo} = 1$, as it is taken as stationary relative to *SORtu*.
The event concentration of *SORtu* relative to *SORtv* consequently is:

$$z_{vu} = \sqrt{1 - \frac{u^2}{c^2}}$$ 13F2

Since one system of reference contains the other and *SORtu* is in motion relative to *SORtv*, with reference to 6N2, the cumulative value of the event concentration of *SORtu*, relative to *SORto*, the stationary system of reference, at a certain time moment of the motion with speed u, is

$$z_u = \sqrt{1 - \frac{v^2}{c^2}} \sqrt{1 - \frac{u^2}{c^2}}$$ 13F3

With reference to 13E5, the transformation process can be written as:

$$\frac{m_o c^2}{z_o} = \frac{m_v c^2}{z_v} = \frac{m_u c^2}{z_u} = ... = \frac{m_{n-1} c^2}{z_{n-1}} = \frac{m_n c^2}{z_n}$$ 13F4

or more precisely:
$$\frac{dm_o c^2}{z_o} = \frac{dm_v c^2}{z_v} = \frac{dm_u c^2}{z_u} = ... = \frac{dm_{n-1} c^2}{z_{n-1}} = \frac{dm_n c^2}{z_n}$$

Where m, m_v, m_u, m_n, m_{n-1} and z_o, z_v, z_u, z_n, z_{n-1} are the mass values and the cumulative (compared to the status of genuine rest) event concentration of the systems of reference at the status of rest and in motion with various speed values. 13F4 connects systems of reference with various time horizons.

It means:
The absolute energy values of the systems of reference of the matter, at different speed levels of the motion and with different time flows within it, are equal:

$$\frac{dm_o c^2}{1} = \frac{dm_o c^2 \sqrt{1 - v^2/c^2}}{\sqrt{1 - v^2/c^2}} = \frac{dm_o c^2 \sqrt{1 - v^2/c^2} \cdot \sqrt{1 - u^2/c^2}}{\sqrt{1 - v^2/c^2} \cdot \sqrt{1 - u^2/c^2}} = ... = \frac{dm_{n-1} c^2}{z_{n-1}} = \frac{dm_n c^2}{z_n}$$ 13F5

The difference in the time flows necessitates using a quotient of the energy intensities and the event concentrations. The event concentration adjusts the time horizon difference and makes it possible to compare energy and work values of different time flows.

The transformation of the mass of the matter into motion determines a certain "direction" of the time flow: it grows (speeds up) from the status at rest to the status of motion:

$$\Delta E = \frac{m_o c^2}{z_o} - \frac{m_v c^2}{z_v} \sqrt{1 - \frac{v^2}{c^2}} \; ; \qquad \Delta t_v = \frac{\Delta t_o}{\sqrt{1 - (v^2/c^2)}}; \quad z_v = \sqrt{1 - \frac{v^2}{c^2}}$$ 13F6

where z_v corresponds to the system of reference in motion at speed v.

The transformation occurs in different time horizons within the systems of reference of the matter, one is part of the other. We do not need to know the reason for or the "trigger" of the transformation, but there is no event and no time to count unless the transformation has happened. Any transformation and motion within the matter means/results in time flow.

There is no motion, meaning no event at the status of *SORto*, when matter is at rest. Any motion within *SORto* means the generation of *SORtv*:

$$\Delta t_v > 0 \qquad \text{means the motion speeds up the time flow.}$$

Once the transformation happens, it relates to the mass value, the energy of which is the drive of the transformation.

S.
13.2.4   *13.2.4. <u>The mass transformation of the matter generates higher energy intensity within the stationary system of reference</u>*

With reference to 13D2, the energy intensity of mass m_o, transformed into motion within *SORto*, the stationary system of reference, is:

13G1
$$\Delta e_{tr} = m_o c^2 - m_o c^2 \sqrt{1 - \frac{v^2}{c^2}} \; ;$$

With reference to the values of the event concentration, the energy intensity of the same transformation within *SORtv*, the system of reference in motion relative to the stationary system of reference, is:

13G2
$$\frac{\Delta e_{tr}}{z_o} = \frac{\Delta e_{v-tr}}{z_v} \qquad \text{where} \quad z_o = 1 \text{ and } z_v = \sqrt{1 - \frac{v^2}{c^2}}$$

13G3
$$\Delta e_{v-tr} = \Delta e_{tr} \sqrt{1 - \frac{v^2}{c^2}} \qquad \text{and} \qquad \Delta e_{v-tr} = m_v c^2 - m_v c^2 \sqrt{1 - \frac{v^2}{c^2}}$$

The absolute energy of the mass transformation is the same. Since the time flow of *SORto* is slower, 13G1 and 13G3 mean that the effect and the *energy intensity of the transformation* within the stationary system of reference are *higher*. The effect is proportional to the value of the event concentration. *The appearance of the energy within systems of reference, distinct in motion, is different.*

S.
13.2.5   *13.2.5. <u>The benefit of the energy intensity</u>*

SORto, the stationary and *SORtv*, the system of reference in motion, represent the systems of reference of the matter with their unique internal balance of mass and energy. But most importantly these systems of reference have their own time flow, the function of the motion of the mass components, demonstrating the unity of the energy, mass, motion and time.

We take that *SORtu* is a system of reference with mass m_u in motion with
Ref $v = u$. The internal energy intensity of the mass transformed into motion in
13G3   this system of reference, in accordance with 13G3, was:

$$\Delta e_{u-tr} = m_u c^2 - m_u c^2 \sqrt{1 - \frac{u^2}{c^2}}$$

13H1

In terms of the "transformed" mass into motion, it would mean:

$$e_{u-tr} = m_{u-tr} c^2$$

13H2

For this case, as we suppose mass m_{u-tr} is in genuine motion, with event concentration of *SORtu* taken as $z_u = 1$

13H3

Consequently, the event concentration of *SORto* would be $z_o = \dfrac{1}{\sqrt{1 - \dfrac{u^2}{c^2}}}$

13H4

The relation of 13H3 and 13H4 would be the same if $z_o = 1$

and $z_u = \sqrt{1 - \dfrac{u^2}{c^2}}$ taken.

With reference to 13G2, the balance of the kinetic energy of the stationary status and the system of reference in motion is:

$$\frac{\Delta e_{o-tr}}{z_o} = \frac{\Delta e_{u-tr}}{z_u}$$

13H5

which, with reference to 13H1, 13H2 and 13H4, gives:

$$e_{o-tr} = \frac{e_{u-tr}}{\sqrt{1 - \dfrac{u^2}{c^2}}} = \frac{m_{u-tr} c^2}{\sqrt{1 - \dfrac{u^2}{c^2}}} = \frac{m_u c^2}{\sqrt{1 - \dfrac{u^2}{c^2}}} - m_u c^2$$

13H6

In accordance with 13H6, mass components in *genuine motion* within the matter (*SORtu*), suddenly part of the stationary status (*SORto*), represent higher energy intensity (the measured value of the energy within the stationary system of reference), inversely proportional to the event concentration of the two statuses.

Because of the difference in the values of the event concentration of the two statuses, the *intensity of the energy generation* (of the slowing down effect of the mass components) within the stationary status is *higher*.

Since the energy intensity within the systems of reference is in fact the "measured" value of the energy, the impact of the energy and work within the systems of reference of "slower time flow" is higher. In order to use this energy effect we have to slow down the mass components or particles in motion within their genuine system of reference and gain the energy benefit of the appearance within the stationary system of reference with higher energy intensity.

S.
13.3

13.3

13.3   **Mass and energy balance of the matter in non adiabatic conditions**

S.

13.3.1   *13.3.1. Non adiabatic system of reference and mass transformation*

What happens if a system of reference, mass value of m_v, transformed from stationary status m_o, is subject to external energy transfer through the boundaries of the system of reference (through the boundaries of the "virtual *non*-adiabatic box")?

The appearance of the energy of the matter within the stationary system of reference is:

13I1

$$e_o = m_o c^2$$

The appearance of the energy of the matter in motion with v:

13I2

$$e_v = m_v c^2 = m_v c^2 - m_v c^2 \sqrt{1 - \frac{v^2}{c^2}} + m_v c^2 \sqrt{1 - \frac{v^2}{c^2}}$$

For simplicity, we suppose that during the external energy supply, while the mass components of the matter are in motion with v, there is no transformation of mass into motion within the system of reference.

External energy through the boundaries of the system of reference will increase the internal energy of the system of reference in motion, but cannot result in transformation of mass into motion.

Ref
S.13.1

As a result of the external energy transfer, the increase of the energy of the system of reference in motion, with reference to Section 13.1, will not change the mass value of the matter in motion. While the speed of the motion of the mass of the matter will be increasing from v, (the speed of the earlier mass transformation) to u, (a higher speed value under the effect of the external energy), the mass of the matter stays constant.

Ref
13.2.2

With reference to Section 13.2.2, the energy balance of the event in different time horizon, corrected by the event concentration, is:

- before the energy transfer to the system of reference in motion with v:

13I3

$$\frac{e_o}{z_o} = \frac{e_v}{z_v}; \qquad \frac{m_o c^2}{z_o} = \frac{m_v c^2}{z_v}$$

- after the energy transfer:

13I4

$$\frac{e_o}{z_o} + \frac{\Delta e_o}{z_o} = \frac{e_v}{z_u}$$

meaning: the energy of the transformation results in speed u of the system of reference in motion and the event concentration is a function of the speed rather than the mass of the matter.

z_v - is the value of the event concentration of the system of reference in motion with v with mass transformation;

z_u - is the event concentration of the same system of reference, with the same transformed mass, but with speed u respectively; and

z_o - is the value of the event concentration of the supposed-to-be stationary system of reference from where the energy is coming.

With reference to 13I4, the increased energy is:

$$\frac{m_o c^2}{z_o} + \frac{\Delta e_o}{z_o} = \frac{m_v c^2}{z_u}$$

1315

$$\frac{m_o c^2}{z_o} + \frac{\Delta e_o}{z_o} = \frac{m_v c^2 - m_v c^2 \sqrt{1 - \left(v^2/c^2 \right)}}{z_u} + \frac{m_v c^2 \sqrt{1 - \left(v^2/c^2 \right)}}{z_u}$$

1316

where:

m_o – is the original mass, within the stationary system of reference, transforming into motion, using its internal energy;

Δe_o – is the external energy intensity through the boundaries of the stationary system of reference, the real appearance of the energy;

$m_v c^2 - m_v c^2 \sqrt{1 - v^2/c^2}$ – is the mass, transformed into motion with speed v, or the kinetic energy of the matter from the point of view of the system of reference in motion;

$m_v \sqrt{1 - \left(v^2/c^2 \right)}$ – is the mass value of the remaining energy reserve at rest of the system of reference in motion with speed v;

$z_u = \sqrt{1 - \left(u^2/c^2 \right)}$ – is the event concentration of the system of reference in motion with increased speed of u, as a consequence of the external energy transfer;

$z_o = 1$ – the event concentration of the stationary system of reference.

The right-hand side of 13I5 and 13I6 is the total energy intensity of the system of reference in motion with transformed mass m_v into motion under the effect of the internal energy. The energy intensity values relate to the event concentration of the motion with u.

Ref
13I5
13I6

There is energy added to the system of reference in motion through the boundaries of the system of reference. With reference to 13I3 and 13I4, the difference between the energies before and after the energy transfer is:

$$\frac{e_o}{z_o} + \frac{\Delta e_o}{z_o} - \frac{e_o}{z_o} = \frac{e_v}{z_u} - \frac{e_v}{z_v} ; \qquad \Delta E_o = \frac{m_v c^2}{\sqrt{1 - \dfrac{u^2}{c^2}}} - \frac{m_v c^2}{\sqrt{1 - \dfrac{v^2}{c^2}}}$$

1317

since $z_o = 1$ is taken: $\Delta e_o = \Delta E_o$

The energy transfer from an external source to *SORtv* the system of reference with mass m_v with speed v, will accelerate *SORtv* into *SORtu*, the system of reference in motion with speed u. There is no transformation of mass during the speeding up from v to u. The increase of the speed is the result of the added external energy.

$u = v$ would mean no energy transfer: $\Delta E = 0$
$u > v$ means external energy is added to the system of reference in motion:
$$\Delta E > 0$$
$u < v$ means energy is taken from the system of reference in motion for external use: $\Delta E < 0$
$u = 0$ is a special case, and it gives a negative value to the work formula. It means the energy of relativistic mass of m_v in motion with v, is taken off and the motion is slowed down to $u=0$.

13J1
$$\Delta E = -\frac{m_v c^2}{\sqrt{1 - \dfrac{v^2}{c^2}}} + m_v c^2$$

Meaning: the loss in speed of the system of reference results in energy generation.

The utilization of external energy changes the energy of the matter and the motion of the matter, but it cannot change the mass. The mass stays constant. The utilisation of the internal energy changes the mass, but it does not change the total energy of the matter.

The energy balance can be addressed from the point of view of different systems of reference. While the absolute energy is equal and the same, the appearance of the energy (the intensity of the energy) differs in systems of reference distinct in motion.

13J1 is equivalent to

13J2
$$\Delta E = -m_o c^2 + m_o c^2 \sqrt{1 - \frac{v^2}{c^2}}$$

Under external accelerating work the mass stays constant and no mass transformation takes place. The internal energy of the matter results in acceleration, in the change of the mass, in the transformation of mass into motion (energy).

14   S.14

The energy quantum

The transformation-acceleration of matter can be divided into an infinite number of stages. Each stage differs from the other by the increased actual speed value of the acceleration. The step by step increase of the speed is taken infinitely small and equal to Δv. We are looking for the value of the transformed mass of the matter at the end of the process.

The first step is the acceleration of *SORto*, the stationary system of reference, into *SORtv*, the system of reference in motion with v. With reference to 13D1, the value of the transformed mass is:

$$\Delta m_{v-tr} = m_o - m_v$$

Ref
13D1

14A1

where m_o - is the mass of the stationary *SORto*, and

m_v - is the mass of *SORtv* in motion.

The next step is the acceleration of the mass from speed v to speed u. It results in the transformation of the mass of the matter:

$$\Delta m_{u-tr} = m_v - m_u \qquad \text{where } m_u \text{ - is the mass of } SORtu.$$

14A2

and so on.

The last step we expect is the acceleration from $i = \lim c$ to c.

$$\Delta m_{c-tr} = m_i - m_c \qquad \text{where } m_i \text{ - is mass of } SORti.$$

$$\Delta v = v - 0 = u - v = \dots = c - i \qquad \lim \Delta v = 0$$

14A3

It is easy to understand the kinetic energy, as far as it is associated with mass. With reference to 14A1, the intensity of the kinetic energy of the acceleration-transformation is:

$$\Delta e_{motion} = \Delta m_{tr} c^2 = m_o c^2 - m_o c^2 \sqrt{1 - \frac{v^2}{c^2}} \; ; \quad \text{or} \quad dm_{tr} c^2 = dm_o c^2 - dm_v c^2$$

14A4

The higher is the speed, the higher is the kinetic energy and the lower is the value of the mass. The energy source of the acceleration-transformation, with reference to Section 13.2.3, is the internal energy of the mass. All stages of the process represent different mass systems of reference at certain speed values of the sphere symmetrical expanding acceleration.

Ref
S.
13.2.3

Ref
13C5
The mass systems of reference in motion have their own time frame, own measured mass, own energy intensity and own event concentration. With reference to 13C5, the absolute energy values are in balance.

Can we imagine that, contrary to our classical views, the kinetic energy in 14A4 is the pure energy status of the matter in transformation?

Ref
S.5
With reference to Section 5, the transformation can be described as

14A5
$$d\left(m_o c^2 - m_v c^2\right) = \frac{dp_{tr}}{dt_{tr}} ds_{tr} ; \quad \text{and} \quad d\left(m_o c^2 - m_v c^2\right) = \frac{d(m_{tr}\upsilon)}{dt_{tr}} ds_{tr}$$

where dp_{tr} – is the momentum of the transformation of the mass;

dt_{tr} and ds_{tr} – are the duration and the path of the mass transformation;

m_{tr} – is the value of the transformed mass; and

υ – is the velocity of the mass transformation, the speed value of the "disappearance" of the matter. It is not equal to v or Δv, the speed or the speed difference between the systems of reference, the different levels of the sphere symmetrical expanding acceleration of the matter.

14A5 means, the mass difference of the two systems of reference transforms into energy. The energy is the matter in motion at speed υ.

Can the value of υ in 14A5 be other than c? Resolving 14A5 we have:

14A6
$$d\left(m_o c^2 - m_v c^2\right) = \left(\frac{dm_{tr}}{dt_{tr}}\upsilon + \frac{d\upsilon}{dt_{tr}}m_{tr}\right)\upsilon dt_{tr}$$

We suppose that the speed of the mass transformation is constant in time:

$$\frac{d\upsilon}{dt_{tr}} = 0 \qquad \text{(The result will prove that this condition is taken correctly.)}$$

Resolving 14A5 we come to:

14A7
$$\frac{d\left(m_o c^2 - m_o c^2 \sqrt{1 - \frac{v^2}{c^2}}\right)}{dm_{tr}} = \upsilon^2 \qquad \text{the integral of which results in:}$$

14A8
$$m_{tr} = m_o \frac{c^2}{\upsilon^2}\left(1 - \sqrt{1 - \frac{v^2}{c^2}}\right)$$

14A7 gives the evidence: there is only one valid option: $\upsilon = c$

Ref
13D1
With reference to 13D1, the value of υ, the speed of the "disappearance" of the mass into energy (motion) *cannot be either more or less than c*, the speed of light, the definition of c at this stage, otherwise the mass balance cannot be maintained:

$$m_{tr} = m_o - m_o \sqrt{1 - v^2/c^2} \qquad \text{and} \qquad m_{tr}c^2 = m_o c^2 - m_o c^2 \sqrt{1 - v^2/c^2} \qquad \text{14A9}$$

We have to note here that the speed of *SORtv*, the system of reference in motion, is still $v < c$, equal to the speed of one of the levels of the step by step sphere symmetrical expanding acceleration of the mass.

We have also to note that we speak about two kinds of transformation:
- the acceleration of the system of reference from one status with certain speed and certain mass, into another with higher speed and less mass; and
- the transformation of the mass into energy with always constant speed of the "mass disappearance" as $\upsilon = c = const$,

At $v = c$, the mass is zero, but the energy exists: *the photon with equal energy quantum* – as result of mass transformation – *has been born.*

We have the *Quantum System of Reference* with no mass, speed of $\upsilon = c = const$ and growing summa energy within our virtual box with unlimited boundaries. And we also have mass systems of reference in transformation, in motion with less and less mass values. The absolute energy of the matter is in balance.

The speed of the quantum system of reference is *c=const*, the time flow indefinable: and the event concentration is zero:

$$t_q = \frac{t_o}{\sqrt{1 - v^2/c^2}}; \qquad \text{at } v = c \qquad z_q = \sqrt{1 - \frac{v^2}{c^2}} \qquad \text{Ref S. 13.2.1}$$

We can divide the acceleration-transformation process of the mass into an infinite number of sub-processes, each of them with infinite short period of time. For each of these infinite short sub-processes, with reference to 14A8, the velocity of the transformation of mass into energy is $\upsilon = c$. It results in Δv speed difference between the mass systems of reference of the matter in motion. This is the transformation from the mass status of the matter at one speed into another with higher speed.

The velocity of the transformation of mass into energy is constant $\upsilon = c$ and stays constant at any mass statuses of the acceleration-transformation process.

Ref 14A6 14A8

With reference to 14A6 and 14A8, $m_o = const$, *c=const*. Either the value of the transforming mass depends on the speed: $m_{tr} = f(v)$; or the actual speed depends on the value of the mass in transformation: $v = f(m_{tr})$.

The matter is genuinely programmed for mass-energy transformation. We can consider that all energy quantum compose the *Quantum System of Reference* with zero event concentration and indefinable time flow, but the quantum status is always part of the mass-energy balance of the matter.

Ref
S.
13.2.1

The step by step increase of the motion of the mass systems of reference for infinite short time periods, with reference to Section 13.2.1, means the sphere symmetrical expanding acceleration of the mass of the matter with $a = dv/dt$ on the "approach" towards the *Quantum System of Reference*.

Each consecutive step of the mass transformation (since the mass transforms on account of its internal mass value), results in less mass value, speeded up time flow and less energy intensity of the transformation.

The *mass transformation* directly produces *photons*. Photons are of equal energy quantum value, creating the *Quantum System of Reference*, a homogenous and stable-in-speed energy field, with an infinite number of photons. The *transformation of systems of reference* produces other systems of reference with less mass values as approaching the *Quantum System of Reference*.

For the demonstration of the energy balance we are taking an adiabatic system of reference in its two statuses of transformation in Fig.14.1.

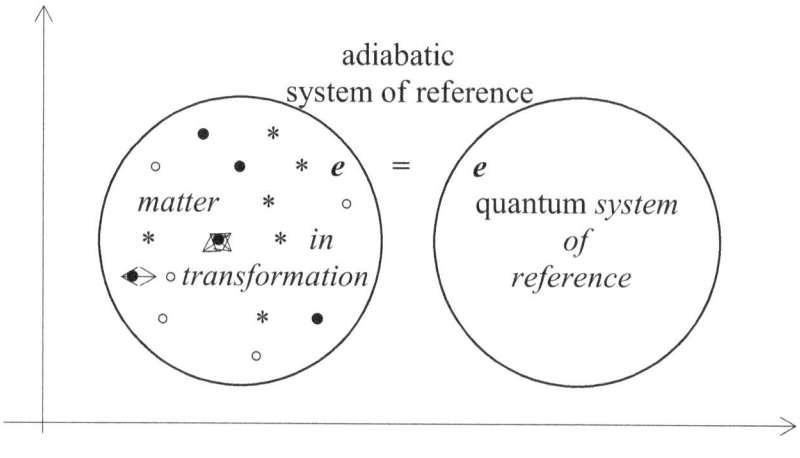

Fig.
14.1

Fig.14.1

The left hand side of Fig.14.1 represents an intermediate stage of the sphere symmetrical expanding acceleration:

mass components in motion, with total energy intensity of $e_v = m_v c^2$

and with mass value of the system of reference:

$$m_v = m_o - m_o\left(1 - \sqrt{1 - (v^2/c^2)}\right) = m_o \sqrt{1 - (v^2/c^2)}$$

The status of the right hand side of the picture is given only for the demonstration of the energy balance, where all mass components are eventually transformed into motion.

> The case is imaginary, because the "boundaries" of the systems of reference are in fact unlimited and because there cannot be a "pure" mass or a "pure" quantum system of reference. We are going to return to this at a later stage.

The last mass status of matter during the transformation of systems of reference is at speed $v = i = \lim at = c$, when the mass value of the matter, the difference of the stationary and the transformed masses, is:

$$m_i = m_o - m_{tr} = m_o - m_o \left(1 - \sqrt{1 - \frac{i^2}{c^2}} \right) \qquad \text{14A10}$$

$$m_i = \lim_{i \to c} m_o \left[1 - \left(1 - \sqrt{1 - \frac{i^2}{c^2}} \right) \right] = 0$$

With reference to Section 7, the balance between the *photons* and the sphere symmetrical expanding acceleration of the *mass* results in sphere symmetrical expanding acceleration for infinite time, the motion with $i = \lim a\Delta t = c$.

What is the energy balance between the two statuses of the *adiabatic* system of reference in Fig.14.1? The energy of the matter during the transformation cannot be lost. It remains within "the adiabatic box" and constitutes the clear energy of the *Quantum System of Reference*, the sum of the energy quantum of an infinite number of photons with speed $v = c$ and *zero* mass.

$$e_{quantum\ SOR} = \sum q$$

where *q=const* denotes the energy of a single quantum

$$e_{vMotion} = e_{quantum\ SOR}$$

With reference to Section 2.1.3, since any vector components of the motion with $v = c$ are equal with c, (we can establish that at this stage of our assessment) the *Quantum System of Reference* is a system of reference with photons in motion with $v = c$ in all directions.

With the progress of the transformation, the value of the energy of the motion, the "clear energy without mass", is growing at each stage of the transformation. The value of the remaining energy reserve at rest is going to be less and less. A larger mass value results in a larger number of photons of equal energy quantum.

14.1
The intensity of the impact of the energy quantum

There is a clear difference between the transformation of the mass and the acceleration of the systems of reference. The first always ends with speed $\upsilon = c$, the second with speed $v = i = \lim a\Delta t = c$, the speed of the motion of the acceleration for infinite time.

The mass transformation of matter produces an infinite number of photons with equal energy *quantum* of $e = q$. While the energy of the mass of the transformation drives the transformation of mass systems of reference forward and the sphere symmetrical expanding acceleration of the mass continues, the mass particles in acceleration collide with the *photons* of the *Quantum System of Reference*. The collision of photons with mass particles of the matter in sphere symmetrical expanding acceleration results in the decrease of the accelerating energy of the mass. At $i = \lim at = c$ the permanent collision and the acceleration will be in balance and the speed of the mass particles in acceleration stays constant.

With reference to Section 5, the loss of the accelerating energy of the mass in motion with $i = \lim a\Delta t = c$ in collision with the photons is:

14B1
$$\frac{de}{dm} = c^2 \lim_{i \to c} \left(1 - \sqrt{1 - \frac{(c-i)^2}{c^2}} \right)$$

dm – is the mass of a single particle or the system of reference of all those particles of the matter in transformation, which are in sphere symmetrical expanding acceleration on the approach to $v = c$, in motion with $i = \lim a\Delta t = c$;

de – is the energy, which would accelerate this particular mass value from speed i up to speed c;

$(c-i)$ – is the speed difference for reaching the quantum stage. In accelerating terms: $adt = dv = (c-i)$

The sphere symmetrical expanding acceleration, the transformation of mass systems of reference, is stopped by the energy quantum of the photons of the *Quantum System of Reference* at $i = \lim a\Delta t = c = const$. In this case, the energy "taken off" by photons must be in balance with the accelerating energy (the accelerating work) of the matter.

Ref
13F4

With reference to 13F4, the correct energy relation between two systems of reference of different time flows must be maintained through the use of event concentration:

14B2
$$\frac{e}{z} = \frac{e_i}{z_i} = \dots = \frac{e_x}{z_x}$$

In this particular case the balance between mass systems of reference of the matter in motion with $i = \lim a\Delta t = c$, the sphere symmetrical expanding acceleration for infinite time and the *Quantum System of Reference*, would be:

$$\frac{de}{z_i} = \frac{de_q}{z_c}; \quad \frac{de_q}{z_c} \text{ however, has no meaning, since the event concentration} \quad \text{14B3}$$

of the quantum system of reference is zero: $z_{v=c} = \sqrt{1 - \frac{v^2}{c^2}} = 0$ 14B4

This is not about the limited validity of 14B2, rather of the fact what 14B3 really means: *there is no pure quantum system of reference on its own.* The photons represent the *kinetic* energy of the matter, part of all systems of reference (in motion).

The photons are products of the mass transformation. They keep the balance with mass systems of reference in sphere symmetrical expanding acceleration. The number of photons grows, but the energy of each energy quantum stays constant: $q = const$

The energy of each photon is not just constant but also must be equal to each other. Why? Because

-   should we find a single photon with higher energy than the energy of the others, the smallest existing energy quantum, it would mean that the transformation of a certain mass particle of the matter resulted in extra energy, and there are photons around with higher energy than the energy of the mass particles of the matter still in acceleration.
-   should we find a single photon with less energy quantum than the others, it would mean that the transformation of the matter into energy ended at a lower quantum energy level and there are mass particles around with less energy than the energy quantum of the photons.

Neither of the two cases above can be valid. Photons must be of equal energy the smallest possible.

What is the work like of the slowing down effect in balance with the sphere symmetrical expanding acceleration of mass systems of reference for infinite time, when photons do not participate in energy exchange?

We write the work formula of the slowing down of the mass particles in 14B1 in a form:

$$\frac{dx}{c^2 dm} = \lim_{i \to c}\left(1 - \sqrt{1 - \frac{(c-i)^2}{c^2}}\right); \quad \frac{qdn}{c^2 dm} = \lim_{(i \to c)}\left(1 - \sqrt{1 - \frac{(c-i)^2}{c^2}}\right) \quad \text{14B5}$$

where $dx = dn \cdot q$ means the summarized energy of the photons and n is an integer. The energy of $(1, 2, \dots n)$ number of photons is in energy balance with the accelerating energy of the mass particles of the system of reference in acceleration for infinite time, the motion with $i = \lim a\Delta t = c$.

Ref
S.6

With reference to Section 6, we have to remember here, that the appearance of events within systems of reference depends on the time flow, the consequence of the velocity of the system of reference. We observe and measure in fact the appearance of events in systems of reference in motion, work and energy intensities rather than their absolute values.

We take in 14B5, that the event, the slowing down of a mass particle, needs n photons to balance the work and needs *time* to happen.

The slowing down happens in two systems of reference at least:
- within *SORti*, the system of reference of the mass particles in sphere symmetrical expanding acceleration, in motion with $i = \lim a\Delta t = c$, and
- within *SORq* the *Quantum System of Reference*, with an infinite number of photons, with equal and constant energy quantum, without energy exchange between photons.

Ref
6K2
6K3
6K4
For good measure and for the sake of the justification we take it that the event also happens within *SORto*, the stationary system of reference at rest with $v=0$.

With reference to 6K2, 6K3 and 6K4, the work and energy intensity values of an event, in general, are:

14C1   - within *SORti*: $w_i = \dfrac{dW}{dt_i}$;   and $e_i = \dfrac{dE}{dt_i}$

14C2   - within *SORq*: $w_c = \dfrac{dW}{dt_c} = 0$;   and $e_c = \dfrac{dE}{dt_c} = 0$

14C3   - within *SORto*: $w_o = \dfrac{dW}{dt_o}$;   and $e_o = \dfrac{dE}{dt_o}$

Within 14C2 the work and energy intensities of the quantum system of reference are zero, where $t_c = \dfrac{t_o}{\sqrt{1-(c^2/c^2)}}$ and while it has no meaning as such, substituting it into 14C2 gives zero intensity result, proving additionally that there is no work and energy exchange between photons. Those are the result of the mass transformation.

Ref
S.6

With reference to Section 6, the absolute value of work and energy of one and the same event, simultaneously happening in two or more systems of reference, distinct in motion, are equal. Its work and energy intensity values are different and are function of the time "flow".

We can write optionally 14B5 as:

14C4
14C5
$$\frac{n}{dt_i} = \frac{W}{q \cdot dt_i} ;\quad \text{or} \quad f = \frac{mc^2}{q}\left(1 - \sqrt{1 - \frac{(c-i)^2}{c^2}}\right)$$

where:
  - W is the absolute value of the accelerating work;

- $E_q = q \cdot n$ is the absolute energy of the photons;
- n is the number of the photons; and
- dt_i is the unit time period of the system of reference in motion with $i = \lim a\Delta t = c$, the acceleration for infinite time.

(With reference to Section 6, the intensities relate to the whole system of reference, therefore, the time reference may not be indicated in the formulas.)

Photons do not participate in the energy exchange. They do not absorb any energy, but their number in the collision, participating in the process, must be in balance with the work or energy values of the event. For balancing one and the same absolute work of an example [$dW_{example}$] within a stationary system of reference (*SORto*) or in any other in motion with v (*SORtv*), the necessary number of photons of the balance is:

$$n = \frac{dW_{example}}{q}$$
14D1

In intensity terms,

for *SORto* it means: and for *SORtv* in motion:

$$\frac{dW_{example}}{q \cdot dt_o} = \frac{dw_o}{q} = \frac{n}{dt_o} = f_o;$$ $$\frac{dW_{example}}{q \cdot dt_v} = \frac{dw_v}{q} = \frac{n}{dt_v} = f_v$$ 14D2
14D3

In 14D2 and 14D3

- dw_o and dw_v are the work intensities of the stationary and the system of reference in motion;

- f_o and f_v denote the *intensity* of the effect of an equal number of photons of the event within the systems of reference accordingly. They obviously indicate and characterize the appearance or impact of the photons of the event for a unit time period within the system of reference.

Since $t_v = \frac{t_o}{\sqrt{1-(v^2/c^2)}}$, the intensities of the impact of the photons in the systems of reference are different:

$$f_v = \frac{n}{dt_o}\sqrt{1-\frac{v^2}{c^2}} = f_o\sqrt{1-\frac{v^2}{c^2}}$$
14D4

as are the intensities (the appearance) of works in as well.

The lower the speed of the motion of the system of reference is, the higher is the impact, the intensity of the participation of the photons within the event.

Ref
14C4
14C5

The message of 14C4 and 14C5 is: for slowing down the motion of a unified mass particle, within *SORti*, the system of reference in sphere symmetrical expanding acceleration, the quantum energy of n photons is necessary. The process happens for dt_i time within *SORti*, the intensity, the impact of the photons for unit time period is f.

Are the photons acting at the unified mass particle in sequence or in parallel, all together?

- Should the n photon be acting in sequence, each for (dt_i/n) time, the full duration of the process would be: $n \cdot (dt_i/n) = dt_i$, as in the real case.

- Should the n photons be acting, on the contrary, for dt_i in parallel, since the energy quantum cannot be divided and the number of photons shall be real and integer, it would result in ndt_i process time. But the duration of the process is only dt_i.

Therefore, we can state in general: the *photons act in sequence*.

We call the intensity of the impact of the photons: *frequency* $= f$.

The dimension of the frequency, in a system of reference with a *second* unit time measurement, is:

14D5
$$f = \left[\frac{1}{\sec}\right] = \frac{1}{\Delta t}$$

Due to acting in energy quantum and in sequence, the effect of the photons has *wave character*, with a *wave length* between two impulses of the photons, denoted as:

14D6
$$\lambda = c \cdot \Delta t = \frac{c}{f}$$

The impact of the photons in interactions depends on their number in the energy balance. The number of the photons must match the absolute work value or the energy difference of the event.

14D7
$$W = \Delta E = n \cdot q$$

The sequence of the impact of the photons in systems of reference with different speed values of the motion depends on the time flow of the system of reference, the *frequency* of the process. Categories *frequency* and *wave length* are indicators of events within systems of reference and are different system by system.

With the definition of the frequency and the energy quantum, we can give the definition of the *duration* of an event within a system of reference in motion with v:

14D8
$$\frac{dW}{dt_v} = \frac{dn}{dt_v}q \quad \text{which gives:} \quad dt_v = \frac{q}{w_v}dn \quad \text{and} \quad \Delta t_v = n\frac{q}{w_v}$$

The genuine dimension of the energy intensity is *energy/time*, the result is as expected.

The duration of the appearance of an event within a system of reference is equal to the total quantum energy of the event divided by the energy intensity difference or the work intensity of the event within the given system of reference.

14.2
Entropy of the mass transformation, the *quantum entropy*

With reference to 13D1, will the transformation of the mass from its stationary status result in the transformation of the full mass into energy?

$$m_{tr} = m - m\sqrt{1 - \frac{v^2}{c^2}}$$

In accordance with 13D1, the substitution of c into v in the formula seems to be giving the transformation of the full mass. The natural *entropy* of any self transformation, however, as also predicted by 14B3, prevents turning the entire energy from one status into another for the balance of the same energy of the matter. The *entropy* of the transformation limits the process. The last portion of the mass cannot be transformed by itself into quantum energy. But what is the value of the remaining mass?

For the definition of the meaning of this *entropy* we take the transformation of mass m from its stationary status until $v = c$. For simplicity we divide the process into two parts: (1) transformation-acceleration until $i = \lim a\Delta t = c$ and (2) further transformation-acceleration up to $v = c$.

With reference to 13D1 and 14A1, the value of the mass transformation, (the energy of which accelerates *SORto* into *SORti* with $i = \lim a\Delta t = c$) is:

$$m_{tr(o-i)} = m - m\sqrt{1 - \frac{i^2}{c^2}} \qquad \text{or} \qquad m_{tr(o-i)} = m - m_i$$

$i = \lim a\Delta t = c$ may be the final mass status of the matter, but it is not the quantum energy status.

With reference to 14A8, the transformation of mass $m_{tr(o-i)}$ ends up at $v = c$. *SORti*, the system of reference in transformation-acceleration is reaching $v = i = \lim a\Delta t = c$.

$$m_{tr(o-i)} = m\frac{c^2}{c^2}\left(1 - \sqrt{1 - \frac{i^2}{c^2}}\right)$$

The meaning of the entropy is that mass systems of reference in transformation cannot reach the quantum status $v = i \neq c$. The energy of the existing photons prevents the transformation of the last portion of the mass.

The sphere symmetrical expanding acceleration of the mass will not stop and the remaining "energy" of the mass in transformation tries to accelerate *SORti*, the system of reference in motion with $i = \lim a\Delta t = c$ up to $v = c$, to reach the speed of the *Quantum System of Reference*. This would need mass energy of:

$$m_{tr(i-c)} = m_i - m_i\sqrt{1 - \frac{(c-i)^2}{c^2}} \qquad \text{where} \qquad v = a\Delta t = c - i$$

Without writing down the formula it is clear that the result will not be giving the transformation of the full mass. The remaining non-transferable mass value will be different than zero. We are now getting to this remaining mass value, deducting all transformed mass values of the transformation from the original mass of the matter. It gives

14E4

$$m_{rem} = m - (m - m_i) - \left(m_i - m_i \sqrt{1 - \frac{(c-i)^2}{c^2}} \right) = m_i \sqrt{1 - \frac{(c-i)^2}{c}}$$

Ref
14.1

With reference to Section 14.1, the work and energy values of the transformed and remaining masses in 14E3 and 14E4 are in line with the intensity of the energy of the photons in balance with the mass transformation. At the same time 14E4 also shows the principle of the entropy: Even without calculating the energy intensity needs of the collision of the transforming mass with the photons, it is seen that the full mass cannot be transformed into energy quantum.

S.
14.2.1

14.2.1. The energy balance of the quantum entropy

For the calculation of the energy balance of the remaining non-transferable mass value, we have to remember that the energy and work values of systems of reference are in fact intensities and relate to different time horizons.

We have to find the correct energy balance between the systems of reference in stationary status *(SORto)*, in motion with $i = \lim a\Delta t = c$ *(SORti)* and in the approach of the quantum system of reference *(SORt(c-i))*. Therefore, the energy intensities must be adjusted by the values of the event concentration. We will have here two choices: (1) we can take *SORti* as stationary relative to *SORt(c-i)*, the approaching system of reference, or (2) relate all values of the event concentration to a stationary system of reference, taken for this purpose.

In the first case the energy balance relation, strictly concerning only for *SORti* and *SORt(c-i)* is:

14F1

$$\frac{m_{rem}c^2}{z^*_{(c-i)}} = \frac{m_i c^2}{z^*_i}$$

where for *only* this particular case $z^*_i = 1$ and $z^*_{(c-i)} = \sqrt{1 - (c-i)^2/c^2}$

The other option is the uniform case, when the values of the event concentration relate to a stationary system of reference. This is more complex, but gives the results in a more visible way.

The relation of the quotients of energy intensities and event concentrations for *SORt(c-i)*, *SORti* and *SORto* is:

$$\frac{m_{rem}c^2}{z_{(c-i)}} = \frac{m_{rem}c^2}{z_i\sqrt{1-(c-i)^2/c^2}} = \frac{m_i c^2}{z_i} = \frac{mc^2}{z_o} \qquad \text{14F2}$$

The energy balance of the transformation from $v = 0$ to c is:

$$\frac{mc^2}{z_o} = \frac{m_{rem}c^2}{z_{(c-i)}}\left(1 - \sqrt{1 - \frac{(c-i)^2}{c^2}}\right) + \frac{m_{rem}c^2}{z_{(c-i)}}\sqrt{1 - \frac{(c-i)^2}{c^2}} \qquad \text{14G1}$$

where $z_{(c-i)} = z_i\sqrt{1-(c-i)^2/c^2}$ is the event concentration of $SORt(c\text{-}i)$, the system of reference in acceleration from i to c, in speed difference $v = c - i$ relative to $SORti$; $z_i = \sqrt{1 - i^2/c^2}$ is the event concentration of $SORti$ and $z_o = 1$, the event concentration of the stationary system of reference.

The components of 14G1 are:

$$\frac{m_{rem}c^2}{z_{(c-i)}}\left(1 - \sqrt{1 - \frac{(c-i)^2}{c^2}}\right); \qquad \text{and} \qquad m_{rem} = m\sqrt{1-i^2/c^2}\cdot\sqrt{1-(c-i)^2/c^2} \qquad \text{14G2}$$

- is the work of the mass from its stationary system of reference to reach the quantum entropy status, in balance with the photons of the *Quantum System of Reference*. The value of the mass, as its appearance is m_{rem}, the work in absolute terms is equivalent to

$$W_{o-q} = mc^2\left(1 - \sqrt{1 - \frac{(c-i)^2}{c^2}}\right) \qquad \text{14G3}$$

$$\frac{m_{rem}c^2}{z_{(c-i)}}\sqrt{1 - \frac{(c-i)^2}{c^2}}; \qquad \text{and} \qquad m_{rem} = m\sqrt{1-i^2/c^2}\cdot\sqrt{1-(c-i)^2/c^2} \qquad \text{14G4}$$

- is the energy of the remaining mass of the matter, which cannot be transformed into motion (into the *Quantum System of Reference*); in absolute terms is equivalent to

$$E_q = mc^2\sqrt{1 - \frac{(c-i)^2}{c^2}} \qquad \text{14G5}$$

(The capital letters mean the energy values are expressed in absolute terms.)

The energy intensity of the remaining mass of the matter within $SORti$, the system of reference in motion with $\lim i = c$, and its quantum energy equivalent are:

$$e_{qi} = \frac{dE_q}{dt_i} = \frac{dn}{dt_i}q = m_i c^2\sqrt{1 - \frac{(c-i)^2}{c^2}} = m_i c^2\sqrt{1 - \frac{a^2\Delta t^2}{c^2}} \qquad \text{14H1}$$

since $dt_i = \dfrac{dt_o}{\sqrt{1-i^2/c^2}}$; and taken as $dt_o = 1$ $\qquad m_i = m\sqrt{1-i^2/c^2}$

We have to note in 14H1 that the intensity and the quantum energy equivalent are related to the system of reference in motion with lim $i= c$ rather than *SORt(c-i)*. *SORt(c-i)* was merely used for identifying the remaining mass. The transformation of the last portion of the mass, with reference to 14E3, is in balance with the energy of *SORti*.

The equivalent mass energy intensity value in 14H1 well correlates to 14E3 and is equal to 14E4, the mass value of the matter in its last non-transferable status as the result of the transformation of the *SORti*.

Since the transformation covers systems of reference with different time horizons, the transformation of mass into energy can only be addressed through the correction of the energy intensities by the values of the event concentration. The correction, in fact, gives absolute values.

14H2
$$\Delta E_{tr} = \frac{mc^2}{z_o} - \frac{e_{qi}}{z_i} = mc^2 - mc^2\sqrt{1 - \frac{(c-i)^2}{c^2}}$$

From 14H2 at $c = const$ and $i = const$ follows that this energy component cannot be decreased. With reference to Section 7.2 on the changing character of a, the acceleration at speed $i = \lim a\Delta t = c$ continues with the time flow growing $\Delta t_{n+1} > \Delta t_n$ and even $a_{n+1} < a_n$ and $F_{n+1} < F_n$

until $\dfrac{dF}{da} \geq m_i \sqrt{1 - \dfrac{i^2}{c^2}}$, since $i = const$

Once however $\dfrac{dF}{da} < m_i \sqrt{1 - \dfrac{i^2}{c^2}}$ the acceleration is out of sequence and the expanding process slows down.

The energy of the generated photons, the transformed mass equivalent prevents transforming the entire mass into motion. We can call this *quantum entropy* or *entropy of the mass transformation* at $i = \lim a\Delta t = c$:

14I1
$$s_i = m_i c^2 \sqrt{1 - \frac{(c-i)^2}{c^2}} = e_{qi}$$

The value of the quantum entropy in 14I1 cannot be decreased: the energy of the mass of the system of reference of the matter in motion with $\lim i = c$ is not sufficient for any further transformation.

The absolute value of the quantum entropy also can be defined. It, however, should be related to the system of reference in motion with lim $i=c$ and would give the smallest value equal to the intensity value in 14I1. In all other possible systems of reference with speed $v<i$ the value of the quantum entropy is higher.

15

Frequency and the quantum balance

$\mathbf{T}$he frequency is the measurement of the sequence of the impact of the photons: the number of the impulses at the detectors of the measurement for a unit period of time. The pulsation of the measurement is the effect of the photons in collision with the detectors. The higher the frequency is, the higher is the number (intensity) of the impact.

Were there any energy exchange between the photons, the work formula of the exchange, with reference to 14C5, would have to be written as

Ref
14C5

$$q\frac{dn}{de} = \left(1 - \sqrt{1 - \frac{\left(c_{q1} - c_{q2}\right)^2}{c^2}}\right); \qquad de = \frac{dE}{dt_i} = m_i c^2 \qquad \text{and}$$

15A1

would mean a speed difference, value of $\left(c_{q1} - c_{q2}\right)$, between the photons.

We may, of course, turn this statement on its head, saying that should there be any speed difference between photons this would result in energy exchange between them. With reference to the sphere symmetrical expanding acceleration of the mass of the matter, the generation of photons, the smallest energy quantum, we see that this statement in both its forms makes no sense. The formula given in 15A1, with speed difference and at the same time with *constant energy quantum*, has no meaning.

Let us rearrange 15A1 and divide both sides of the equation by

$$\sqrt{1 - \frac{\left|c_{q1} - c_{q2}\right|^2}{c^2}}$$

15A2

The result is the relativistic *work intensity* formula, the one Einstein introduces as the work formula replacing the classical Newtonian.

Ref
S.5.02.
b

(With reference to Section 5.02.b, we know that this is not a "replacement", rather the result of the classical approach, since the basis for both, the relativistic and the classical formula, is the same.)

The resulting expression is $\dfrac{q \cdot dn}{\sqrt{1 - \dfrac{\left|c_{q1} - c_{q2}\right|^2}{c^2}}} = \dfrac{de}{\sqrt{1 - \dfrac{\left|c_{q1} - c_{q2}\right|^2}{c^2}}} - de$ from which,

15A3

with reference to 15A1, it equally follows that

$$q \frac{dn}{de} = 0, \qquad q = const \text{ means } c_{q1} = c_{q2}$$

Max Planck has given the value of the energy quantum. He has found the correct dimension, the one corresponding to the system of reference of the measurement, the system of reference on the surface of the *Earth*. With reference to 14C5, and to the fact that the value of the work within 14C5 is understood in its absolute value,

Ref
14C5
15A4

$$df_i = \frac{n}{dt_i} = \frac{c^2 dm \cdot dt_i}{q \cdot dt_i} \left(1 - \sqrt{1 - \frac{(c-i)^2}{c^2}} \right)$$

the dimension, as it indeed is, must be [*Joule·sec*]. The measurement gives naturally the product of $(q \cdot dt_i)$ as a combined value:

$$(q \cdot dt_i) = H = 6.626 \cdot 10^{-36} [Joule·sec]$$

At frequency $f = 1$, $(q \cdot dt_i)$ provides the energy of a single quantum:

15A5

$$q = (q \cdot dt_i) \frac{1}{dt_i} = q \qquad \text{since} \quad \left[f = \frac{1}{dt_i} \right]$$

The Planck constant perfectly matches the experimental results. It is valid however (only) to the [time] system of reference of the *Earth*. For any other system of reference it must be changed, depending on the motion (and the time flow) of the system of reference of the measurement.

S.
15.1

15.1.
The sequence of the *blue* and *red shifts*

With reference to Section 9, the *blue shift* is the increase of the frequency of electromagnetic waves in collision with systems of reference in acceleration. (Originally the shift marks the Pound-Rebka-Snider experiment at Harvard University, on the increase of the frequency of electromagnetic waves under the effect of Gravitation.) Section 9 proves that the collision of electromagnetic waves with systems of reference or inert bodies in acceleration result in the same effect. Here we are giving the quantum balance explanation of the shift.

We are taking system of reference *SORti* in motion with $i = \lim a\Delta t = c$. *SORti* represents a system of reference in sphere symmetrical expanding acceleration for infinite time.

Let us assume a structure with a photon Emitter and a photon Detector built at level $h=22.5\ m$ above the surface of the system of reference. The Emitter impacts, the Reflector at the surface of system of reference (of the *Earth*) re-impacts (reflects back), the Detector detects the impact (of the photons). We analyze the energy balance of the collision at the Reflector on the surface of the system of reference, at the Emitter and at the Detector at level *h*. The case is presented in Fig.15.1.

S.
15.1.1

15.1.1. The frequency of the impact

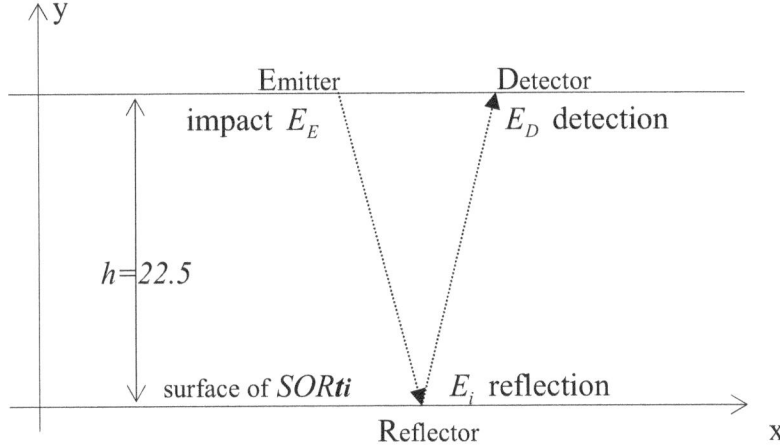

Fig.15.1

Fig
15.1

The absolute energy of the impacted photons at level *h* is equal to the difference of (the energy) of the mass, utilised for the impact. It is:

$$E_E = (M - m)c^2 = n \cdot q \qquad \text{15B1}$$

M – is the mass of the *SORti* before the photon impact;
m – is the mass of *SORti* after the photon impact;
n – is the number of photons of equal energy quantum of *q*.

The energy intensity of the photon impact within *SORti* is:

$$e_E = \frac{d(M - m)}{dt_i}c^2 = \frac{n}{dt_i}q \qquad \text{or} \qquad e_E = f_i \cdot q \qquad \text{15B2}$$

dt_i – determines how many photons have been impacted for a unit period of time. It characterises the intensity of the impact.

The energy balance without the acceleration of *SORti* would be an easy case: the energy of the reduced mass of *SORti* together with the energy of the photons would give the original energy of *SORti*:

In intensity terms: $\dfrac{dE_M}{dt_i} = \dfrac{dE_m + dE_E}{dt_i}$ 15B3

E_M – is the energy of *SORti* before the photon impact;

E_m – is the energy of *SORti* after the photon impact.

But *SORti* is accelerating, which generates additional energy capability.

The work intensities of the acceleration of *SORti*, before and after the photons were impacted are:

15B4
15B5

$$\text{before:} \quad \frac{dW_M}{dt_i} = \frac{dM}{dt_i}c^2\left(1 - \sqrt{1 - \frac{(c-i)^2}{c^2}}\right) \quad \text{after:} \quad \frac{dW_m}{dt_i} = \frac{dm}{dt_i}c^2\left(1 - \sqrt{1 - \frac{(c-i)^2}{c^2}}\right)$$

As result of the portion of the mass, used for the photon impact, *SORti* has an accelerating work surplus of:

15B6

$$W = (M - m)c^2\left[1 - \sqrt{1 - \frac{v^2}{c^2}}\right];$$

Ref
3C3

The surplus in 15B6 is gathered by *SORti* at the moment of the impact of the photons, since, with reference to 3C3, the location of the emitter is at distance

$$h = \frac{c^2}{a}\left(1 - \sqrt{1 - \frac{v^2}{c^2}}\right)$$ "ahead" on the accelerating path of *SORti,*

$v = a\Delta t = (c - i)$ is calculated as the result of the actual acceleration: $a = dv/dt$

The work surplus of *SORti* in 15B6 in intensity terms is:

15B7

$$\frac{dW}{dt_i} = \frac{d(M-m)c^2}{dt_i}\left[1 - \sqrt{1 - \frac{v^2}{c^2}}\right]$$

S.
15.1.2   *15.1.2. The frequency at the reflector*

The absolute energy of the impacted photons at the collision with the Reflector is equal to their absolute energy at the Emitter:

$$E_R = E_E$$

The energy balance of the collision at the Reflector is determined by the time relations. It establishes the intensity of the energy exchange and the frequency of the photons.

15C1

$$e_R = \frac{dE_R}{dt_x} = \frac{d(M-m)}{dt_x}c^2 = \frac{dn}{dt_x}q$$

e_R – is the energy intensity of the reflection;

dt_x – is the "time count" in the system of reference of the reflection, result of the effect of the collision.

The original energy balance at the Emitter is:

15C2

$$\frac{dE_M}{dt_i} = \frac{dE_m}{dt_i} - \frac{dW}{dt_i} + \frac{dE_E}{dt_i}$$

The components of the energy balance are:

- $E_M = Mc^2$ is the summarised energy before the impact of the photons;
- $E_m = mc^2$ is the energy of *SORti* with its reduced mass, after the impact;
- W is the energy surplus of the acceleration of *SORti*;
- E_E is the energy of the photons impacted.

The description of the balance at the Reflector is:

$$\frac{dE_M}{z_x dt_i} = \frac{dE_m}{z_x dt_i} - \frac{dW}{z_x dt_i} + \frac{dE_R}{x_x dt_x}; \qquad \text{15C3}$$

- $dE_M/dt_i = e_M$ - is the intensity of the energy of the full mass of *SORti*, before the impact of the photons;

- $dE_m/dt_i = e_m$ - is the intensity of the energy of the reduced mass of *SORti* after the photons are impacted;

- $dE_E/dt_i = e_E$ - is the mass energy equivalent of the impact of the photons at the Emitter (E) at the moment of the impact. It also can be written as $n \cdot q/dt_i$, where the numerator is the product of the number and the (equal) energy quantum of the photons.

- $dW/dt_i = w_i$ - is the work intensity surplus of the acceleration of the full and reduced mass difference of *SORti*. The work surplus means the energy difference between the Emitter and the Reflector in the accelerating path.

- $dE_R/dt_x = e_R$ - is the mass energy equivalent of the reflection (re-impact) of the photons at the Reflector. It also can be written as $n \cdot q/dt_x$, where dt_x marks the unit time of the re-impact at the Reflector, what we are looking for, in order to establish the frequency of the reflection.

- z_x - is the event concentration of *SORtx*, the system of reference of the collision at the Reflector, the use of which needs explanation:

The impact of photons happens in *SORti*, a well-defined system of reference in motion with $i = \lim a\Delta t = c$. The collision at the Reflector is also within *SORti*, but we do not know what would be the time "count" at the reflection. The energy balance of the collision of the photons and *SORti* determines the time consequence (the new time count) of the collision: In fact *SORtx* is a new expression of *SORti*, the system of reference of mass m in motion. The new time count is denoted through t_x, function of the speed of the motion of *SORtx*.

At the same time, with reference to Section 13.3 (in non-adiabatic energy transfer conditions), external energy does not influence the mass value just modifies the speed of the motion of the system of reference, either speeding it up or slowing it down. Therefore, the energy intensity of the mass of *SORtx* remains equal to *SORti*.

The number of the photons in collision must match the absolute energy of the event.

15C3 means:

$$\frac{dMc^2}{z_x dt_i} = \frac{dmc^2}{z_x dt_i} - \frac{d(M-m)}{z_x dt_i}c^2\left(1 - \sqrt{1 - \frac{v^2}{c^2}}\right) + \frac{d(M-m)}{z_x dt_x}c^2 \qquad \text{15C4}$$

And 15C4 gives:

$$\frac{d(M-m)}{dt_i}c^2\left[1 + 1 - \sqrt{1 - \frac{v^2}{c^2}}\right] = \frac{d(M-m)}{dt_x}c^2; \qquad \text{15C5}$$

From 15B2, the energy intensity of the impact is:

15C6
$$\frac{d(M-m)}{dt_i}c^2 = \frac{dn}{dt_i}q$$

The number of the photons at the *re-impact* at the Reflector is the same:

$$d(M-m)c^2 = n \cdot q$$

but the intensity of the event is:

15C7
$$\frac{d(M-m)}{dt_x}c^2 = \frac{dn}{dt_x}q$$

Substituting 15C6 and 15C7 into 15C5, it gives the summarized frequency of the photons after the collision at the Reflector:

15D1
$$\frac{dn}{dt_x} = \frac{dn}{dt_i}\left[1+1-\sqrt{1-\frac{v^2}{c^2}}\right]; \quad \text{or} \quad f_x = f_i\left[1+1-\sqrt{1-\frac{v^2}{c^2}}\right]$$

The increase (*blue shift*) of the frequency at the Reflector is:

15D2
$$\Delta f_x = f_i\left[1-\sqrt{1-\frac{v^2}{c^2}}\right]$$

The time relation of *SORti* before and after the collision with the photons at the Reflector.

15D3
$$dt_x = \frac{dt_i}{1+1-\sqrt{1-\frac{v^2}{c^2}}}; \quad \text{which means} \quad dt_x < dt_i$$

The consequence of the collision is the slowing down of the motion of *SORti*. The slow down of the motion slows down the time flow. The frequency at the Reflector is:

15D4
$$f_R = f_x = f_i\left[1+1-\sqrt{1-\frac{v^2}{c^2}}\right]$$

S.
15.1.3   *15.1.3. The frequency at the Detector*

The original energy balance of the impact of the photons at the Emitter was related to dt_i.

With the slowing down effect at the Reflector, the frequency of the impact is increased. All components of the balance at the Reflector relate to dt_x, the "time count", consequence of the slow down:

15D5
$$\frac{dE_M}{z_x dt_x} = \frac{dE_m}{z_x dt_x} + \frac{dW}{z_x dt_x} + \frac{dE_R}{z_x dt_x}$$

The energy balance of the collision at the Detector (D) has the following components:

$$\frac{dE_M}{z_y dt_x} = \frac{dE_m}{z_y dt_x} + \frac{dW}{z_y dt_x} + \frac{dE_D}{z_y dt_y} \qquad \text{15E1}$$

- $dE_D / dt_y = e_D$ - is the energy intensity of the impact of the photons at the Detector. It also can be written as $n \cdot q / dt_y$, where dt_y is the "time count" at the Detector.

$$e_D = \frac{d(M-m)}{dt_y} c^2 = \frac{dn}{dt_y} q \qquad \text{15E2}$$

- $dW / dt_x = w_x$ - is the work intensity *deficit* of the acceleration of the difference of the full and reduced mass of *SORti*. The work deficit relates to the accelerating path between the Reflector and the Detector.

- z_y - is the value of the event concentration of the system of reference of the collision at the Detector.

15E1 will be in form:

$$\frac{dMc^2}{z_y dt_x} = \frac{dmc^2}{z_y dt_x} + \frac{d(M-m)}{z_y dt_x} c^2 \left(1 - \sqrt{1 - \frac{v^2}{c^2}}\right) + \frac{d(M-m)}{z_y dt_y} c^2 \qquad \text{15E3}$$

and 15E3 gives:

$$\frac{dn}{dt_y} q = \frac{d(M-m)}{dt_x} c^2 \left(1 - 1 + \sqrt{1 - \frac{v^2}{c^2}}\right)$$

The value of the red-shifted frequency is:

$$\frac{dn}{dt_y} = \frac{dn}{dt_x}\left(1 - 1 + \sqrt{1 - \frac{v^2}{c^2}}\right); \qquad \text{from 15C1} \quad \frac{d(M-m)}{dt_x} c^2 = \frac{dn}{dt_x} q;$$

The value of the blue shifted frequency (from 15D1):

$$\frac{dn}{dt_x} = \frac{dn}{dt_i}\left(1 + 1 - \sqrt{1 - \frac{v^2}{c^2}}\right)$$

15E3 gives the time relations of the collision at the Detector, relative to the original:

$$\frac{dt_i}{dt_y} = \sqrt{1 - \frac{v^2}{c^2}}\left(1 + 1 - \sqrt{1 - \frac{v^2}{c^2}}\right) \qquad \text{15E4}$$

15E4 gives the frequency correction at the Detector (*red shift*):

$$f_D = f_y = f_x\sqrt{1 - \frac{v^2}{c^2}} = f_i\sqrt{1 - \frac{v^2}{c^2}}\left(1 + 1 - \sqrt{1 - \frac{v^2}{c^2}}\right) \qquad \text{15E5}$$

The effect of the collision at the Detector is decrease in frequency: $f_D < f_x$ 15E6

It means the impact of the photons in the collision at the Detector "speeds up" the motion of *SORti*.

15E7   The frequency at the Detector is less than the original: $f_D < f_E$

Since, with reference to 15E5 and the deduction in 15E8:

$$\sqrt{1-\frac{v^2}{c^2}}\left(1+1-\sqrt{1-\frac{v^2}{c^2}}\right)<1$$

15E8 $A=\sqrt{1-v^2/c^2};\quad A(2-A)<1;\quad 2A-A^2<1;\quad -1+2A-A^2<0;\qquad -|1-A|^2<0;$

From 15E4 follows, the collision speeds up the time flow:

15F1 $$dt_y=\frac{dt_i}{\sqrt{1-\frac{v^2}{c^2}}\left(1+1-\sqrt{1-\frac{v^2}{c^2}}\right)};\qquad\qquad dt_y>dt_i$$

The energy balance at the collision with the Detector relates to the time system, related to dt_y:

15F2 $$\frac{dE_M}{dt_y}=\frac{dE_m}{dt_y}+\frac{dW}{dt_y}+\frac{dE_D}{dt_y}$$

S.
15.1.4   *15.1.4. The energy benefit of the Blue-Red shift sequence*

From the comparison of 15D3 with 15F2 it follows

the frequency at the Reflector: $f_R > f_E$,

the frequency at the Detector is: $f_D < f_E$

With reference to 15B2:

Ref
15B2 $$e_E=\frac{n}{dt_i}q=f_i\cdot q;\qquad E_E=f_i\cdot q\cdot dt_i$$

With reference to 15C1:

Ref
15C1 $$e_R=\frac{n}{dt_x}q=f_x\cdot q;\qquad E_R=f_x\cdot q\cdot dt_x$$

With reference to 15E2:

Ref
15E2 $$e_D=\frac{dn}{dt_y}q=f_y\cdot q;\qquad E_D=f_y\cdot q\cdot dt_y$$

15G1   The absolute energy values are equal at each collision: $E_E=E_R=E_D$
The difference of the process is in energy intensities.

We can also express the absolute values through the product of the frequency and the Planck-constant. The Planck-constant however is function of the time flow: $H=q\cdot dt$

The absolute energy of the collision at the Emitter is:

15H1 $$E_E=\frac{n}{dt_i}q\cdot dt_i=f_iH_i$$

at the Reflector: at the Detector:

$$E_R = \frac{n}{dt_x} q \cdot dt_x = f_x H_x \qquad\qquad E_D = \frac{n}{dt_y} q \cdot dt_y = f_y H_y$$

15H2
15H3

The relations of the intensities and the event concentrations are equal:

$$\frac{e_E}{z_i} = \frac{e_R}{z_x} = \frac{e_D}{z_y}$$

15H4

With reference to Section 6.1, 6.2 and 15B2 and taking $z_i = 1$ for the event concentration of *SORti,* from 15H4 follows:

$$e_E = \frac{z_i}{z_i} e_E = f_i q$$

15H5

$$e_R = \frac{z_x}{z_i} e_E = \left(2 - \sqrt{1 - \frac{v^2}{c^2}}\right) e_E = \left(2 - \sqrt{1 - \frac{v^2}{c^2}}\right) f_i q \qquad e_R > e_E$$

15H6

$$e_D = \frac{z_y}{z_i} e_E = \sqrt{1 - \frac{v^2}{c^2}}\left(2 - \sqrt{1 - \frac{v^2}{c^2}}\right) e_E = \sqrt{1 - \frac{v^2}{c^2}}\left(2 - \sqrt{1 - \frac{v^2}{c^2}}\right) f_i q$$

15H7

with reference to 15E7: $e_D < e_E$

The energy intensity values now are comparable:

$$\Delta e_R = e_R - e_E = \left(1 - \sqrt{1 - \frac{v^2}{c^2}}\right) f_i q$$

15I1

$$\Delta e_D = e_D - e_E = \left[\sqrt{1 - \frac{v^2}{c^2}}\left(2 - \sqrt{1 - \frac{v^2}{c^2}}\right) - 1\right] f_i q$$

15I2

The energy intensity provided by the photons at the Detector is the difference between the red shift at the Detector and the blue shift at the Reflector:

$$\Delta e_{D-R} = e_D - e_R = -\left(1 - \sqrt{1 - \frac{v^2}{c^2}}\right) \cdot \left(2 - \sqrt{1 - \frac{v^2}{c^2}}\right) f_i q$$

15I3

The negative sign of 15I3 demonstrates the higher value of the red shift.

With reference to 15I2, we can find a different position for the Detector, where the reflected electromagnetic waves can keep the original energy intensity (frequency):

$$\sqrt{1 - \frac{a^2 \Delta t_h^2}{c^2}}\left(2 - \sqrt{1 - \frac{a^2 \Delta t^2}{c^2}}\right) = 1 \quad \text{and} \quad \sqrt{1 - \frac{a^2 \Delta t_h^2}{c^2}} = \frac{1}{2 - \sqrt{1 - \frac{a^2 \Delta t^2}{c^2}}}$$

15J1

In this case we have to suppose, however, that the speed component of the red shift is different, since the accelerating potential of the location of the receptor, with reference to 3C3 is:

$$h_h = \frac{c^2}{a}\left(1 - \sqrt{1 - \frac{a^2 \Delta t_h^{\,2}}{c^2}}\right)$$

And the corrected distance with equal blue and red shifts is:

15J2

$$h_h = \frac{c^2}{a}\left(1 - \frac{1}{2 - \sqrt{1 - \frac{a^2 \Delta t^2}{c^2}}}\right)$$

If we take for simplicity that $a=const$, the modified location of the Detector for the Pound-Rebka-Snider experiment, where the frequency of the photons would be equal to the original, at the induction by the Emitter, is: *22.408 m*, where *a=9.81 m/s; Δt=2.137 sec*. The distance of the original induction is: *22.5 m*.

16

Sphere symmetrical accelerating collapse

With reference to Section 14, in the transformation of the matter from its mass status into energy, there will be a point when the internal energy of mass $m_i\sqrt{1-(c-i)^2/c^2}$ (value m at rest) in sphere symmetrical expanding acceleration, will be not sufficient for accelerating any more against the opposing energy of the *Quantum System of Reference*.

With reference to Section 14.2.1, once $\dfrac{dF}{da} < m_i\sqrt{1-\dfrac{i^2}{c^2}}$,

the slow-down of the sphere symmetrical expanding acceleration starts.

With reference to 14E3, the work intensity of the sphere symmetrical expanding accelerating of the mass – until the slow-down – keeps the balance in collision with n photons of the *Quantum System of Reference*:

$$m_{tr(i-c)}c^2 = m_ic^2 - m_ic^2\sqrt{1-\frac{(c-i)^2}{c^2}} = \frac{dn}{dt_i}q = f\cdot q \qquad \text{16A1}$$

in 16A1 f is the frequency of the collision

The slow-down means:

$$m_ic^2 - m_ic^2\sqrt{1-\frac{(c-i)^2}{c^2}} < f\cdot q \qquad \text{14A2}$$

and the internal energy of the mass cannot keep the motion with speed $i = \lim a\Delta t = c$. As a direct consequence, the *quantum entropy* increases.

The work of the mass transformation is less than the value necessary for reaching the status of the quantum entropy. The slowing down will result in speed v, a value less than $i = \lim a\Delta t = c$, and the *quantum entropy* belonging to the motion with v will be increased. In fact, the *Quantum System of Reference* slows down the acceleration.

$$s_v = \frac{s_i}{z_v} \qquad \text{16A3}$$

As 16A3 shows, the value of the quantum entropy within the system of reference (of the mass) with speed v in slowing down course. z_v the event concentration will be of different, increased value.

With reference to 14I1

$$s_i = m_i c^2 \sqrt{1 - \frac{(c-i)^2}{c^2}} = e_{qi} = f_i \cdot q$$

the value of the quantum entropy in the new circumstances of the slowing down will be:

16A4

$$s_v = \frac{s_i}{z_v} = \frac{m_i c^2 \sqrt{1 - \frac{(c-i)^2}{c^2}}}{\sqrt{1 - \frac{(c-v)^2}{c^2}}} ; \qquad \text{where } z_v = \sqrt{1 - \frac{(c-v)^2}{c^2}}$$

Why do we use in 16A4 $(c-v)$ instead directly using v ?

From the principal relativistic point of view we can equally refer to both, the theoretical stationary system of reference or the *Quantum System of Reference*. Let us try to express the slowing down work relative to the stationary system of reference.

16A5

$$W = mc^2 \left(1 - \sqrt{1 - \frac{[c-(c-v)]^2}{c^2}} \right) = mc^2 \left(1 - \sqrt{1 - \frac{v^2}{c^2}} \right)$$

Using v in 16A4 would mean an accelerating work from zero to speed v. Instead we have a slowing down from i to v, and z_v, the value of the event concentration must address this fact. Its value for the system of reference in slowing down course relative to the *Quantum System of Reference* is:

16A6
16A7

$$z_v = \sqrt{1 - \frac{(c-v)^2}{c^2}} ; \qquad \text{or } z_u = \sqrt{1 - \frac{u^2}{c^2}} ; \qquad \text{where } u = (c-v)$$

The format of the expression of the speed in 16A7 has its significance. It denotes the speed value relative to the *Quantum System of Reference*. It gives correct characterisation of the motion in line with the principles of relativity. Obviously: $z_v = z_u$

For simplicity we will keep the expression through $(c-v)$, the reference to the switch to $u = (c-v)$ will be given with the necessary explanation.

As a consequence of the slowing down, the *quantum entropy* grows:

16A8

$$s_v > s_i$$

With reference to 14H1 and 14I1, from 16A4 it follows that the frequency, equivalent to the quantum entropy is:

16A9

$$f_v = \frac{f_i}{z_v} = \frac{m_i c^2 \sqrt{1 - \frac{(c-i)^2}{c^2}}}{q \sqrt{1 - \frac{(c-v)^2}{c^2}}}$$

With the slowing down, the impact of the photons is growing and the frequency increases. We can also turn the statement round and say: the impact of the growing frequency slows down the motion.

$$f_v > f_i$$

The sphere *symmetrical accelerating collapse* starts.

16.1.
The sphere symmetrical accelerating collapse

S. 16.1

Under "the pressure" of the growing frequency and quantum entropy, the matter with mass of $m_i\sqrt{1-(c-i)^2/c^2}$ is collapsing. With reference to Section 13.1, during the accelerating collapse, as being the effect of the "external energy" of the collision with photons, the mass values remains the same. The work intensity of the impact of the photons is:

Ref S.13.1

$$w_{slow} = \frac{m_i c^2 \sqrt{1-\dfrac{(c-i)^2}{c^2}}}{\sqrt{1-\dfrac{(c-v)^2}{c^2}}} - m_i c^2 \sqrt{1-\frac{(c-i)^2}{c^2}}$$

16B1

The numerator in 16B1 denotes the mass value of the quantum entropy, the denominator indicates the speed difference relative to c.

With reference to 16B1, with the decrease of the speed of the mass, the values of the work intensity and the frequency grow:

Ref 16B1

at $\lim v = 0$

$$\lim w_{v\to 0} = \infty$$

and with reference to 16A7: $\quad f_v = \dfrac{f_i}{\sqrt{1-\dfrac{(c-v)^2}{c^2}}} \qquad \lim f_{v\to 0} = \infty$

Ref 16A7

The accelerating collapse slows down the time flow and at $\lim v = 0$ the measured time approaches the "time measurement" of the stationary system of reference:

$$\lim_{v\to 0} t_v = \frac{1}{f_v} = 0$$

16B2

At the end of the sphere symmetrical accelerating collapse, there will be a moment when the quantum entropy (an intensity category) will be equal to the energy intensity of the matter in its collapsing status:

$$s_{\lim v=0} = e_{\lim v=0}$$

16B3

The energy intensity of the collapsing matter in 16B3 above is identical to the first component of the equation in 16B1:

16B4
$$e_v = \frac{m_i c^2 \sqrt{1 - \frac{(c-i)^2}{c^2}}}{\sqrt{1 - \frac{(c-v)^2}{c^2}}} = s_v$$

At the end of the slowing down process there should be an inflection point before the speed would reach $v=0$. Here the sphere symmetrical collapse turns into sphere symmetrical acceleration. It actually means that the energy of the measured mass will be higher than the impact of the photons. The sphere symmetrical expanding acceleration starts. It cannot be otherwise, since at $v=0$, both the energy intensity of the measured mass and the quantum entropy would have no meaning. Therefore we take for the value of the denominator

16C1
$$\lim_{v \to 0}(c-v) = i$$

As it is for the sphere symmetrical expanding acceleration, so it is for the accelerating collapse: In accordance with the entropy rule, the quantum energy cannot fully transform itself into mass, the other form of the matter.

Ref
S.14   With reference to Section 14, the mass in motion with $i = \lim a\Delta t = c$ is

16C2
$$m_i = m\sqrt{1 - \left(i^2 / c^2\right)}$$

Substituting 16C1 and 16C2 into 16B4, the formula gives the energy of the mass, result of the sphere symmetrical collapse:

16C3
$$\lim_{v \to 0} e_o = mc^2 \sqrt{1 - \frac{(c-i)^2}{c^2}} \qquad \text{(mass entropy)}$$

Ref
S.17   Section 17 further will prove, the remaining quantum energy is

16C4
$$e_q = mc^2 \left(1 - \sqrt{1 - \frac{(c-i)^2}{c^2}}\right)$$

and with that *the energy balance of the matter is granted!*

Are the photons in collision with the mass incorporating during the sphere symmetrical collapse?

The energy balance of the collapse is:

16D1
$$\frac{mc^2 \sqrt{1 - \frac{i^2}{c^2}} \sqrt{1 - \frac{(c-i)^2}{c^2}}}{\sqrt{1 - \frac{(c-v)^2}{c^2}}} - mc^2 \sqrt{1 - \frac{i^2}{c^2}} \sqrt{1 - \frac{(c-i)^2}{c^2}} = m_q c^2$$

$m_q c^2$ - on the right hand side of 16D1 denotes the energy of the photons in collision.

16D1 is the work formula of the collapse under the effect of the "external energy" of the photons. The rearrangement of the sides of the equation gives the increase of the internal energy of the mass:

$$mc^2\sqrt{1-\frac{i^2}{c^2}}\sqrt{1-\frac{(c-i)^2}{c^2}}\left(1-\sqrt{1-\frac{(c-v)^2}{c^2}}\right)=m_qc^2\sqrt{1-\frac{(c-v)^2}{c^2}}$$ 16D2

$m_qc^2\sqrt{1-\frac{(c-v)^2}{c^2}}$ is the energy of the "photons" resulting the collapse   16D3

but there is no photons with mass of $m_q\sqrt{1-\frac{(c-v)^2}{c^2}}$ 16D4

16D4 means, the photons of the collapse will have mass status.

The supposed re-transformation of the energy quantum into measurable mass should be in line with the energy balance of the matter. Since the system is adiabatic and there is no energy additionally added to the process in any form, in order to keep the energy balance, the increase of the mass should also increase v, the actual velocity of the collapse:

$$\frac{m_vc^2\sqrt{1-\frac{(c-i)^2}{c^2}}}{\sqrt{1-\frac{(c-v)^2}{c^2}}}=\frac{m_ic^2\sqrt{1-\frac{(c-i)^2}{c^2}}}{\sqrt{1-\frac{(c-\upsilon)^2}{c^2}}}$$ 16D5

In 16D5

m_i - is the mass of the quantum entropy; m_v - is the mass value with the supposed photons incorporated; υ - with reference to 16A4, it is the speed corresponding to the value of the quantum entropy as a result of the slowing down effect of the collapse; v - is the supposed speed, the result of the photon incorporation.

Because of the incorporation, it is taken that $m_v > m_i$,

the balance in 16D1 needs: $(c-v) < (c-\upsilon)$, which means $v > \upsilon$

And so, in this case, we would have a larger mass at a higher absolute speed value of the collapse, whereas higher mass must be belonging to a lower speed.

If we suppose that as the result of the incorporation and $m_v > m_i$ there is no speed increase and $\upsilon = const$, the internal energy of the collapsing mass would become higher than the quantum entropy:

$$\frac{m_vc^2\sqrt{1-\frac{(c-i)^2}{c^2}}}{\sqrt{1-\frac{(c-\upsilon)^2}{c^2}}}>\frac{m_ic^2\sqrt{1-\frac{(c-i)^2}{c^2}}}{\sqrt{1-\frac{(c-\upsilon)^2}{c^2}}}$$ 16D6

Answering the question on the incorporation of photons, based on 16D4 we shall suppose *yes*, but based on 16D6 we shall suppose *no*. It consequently means the uncertainty of the event. The final answer to the question will be given in Section 18.2.

16.2
Time and space balance

With the sphere symmetrical expanding acceleration and with the sphere symmetrical accelerating collapse we are getting to a certain pulsation of the matter, which leads to the two ends of its existence. At one end to the quantum entropy, with reference to 14I1, equal to

$$s_i = m_ic^2\sqrt{1-\frac{(c-i)^2}{c^2}}=e_{qi}=f_i\cdot q$$

and at the other end, with reference to 16B4 and 16C3 to the "mass entropy" , energy intensity, equal to

Ref
16B4
16C3

$$e_v = \frac{m_i c^2 \sqrt{1 - \dfrac{(c-i)^2}{c^2}}}{\sqrt{1 - \dfrac{(c-v)^2}{c^2}}} = s_v = f_v \cdot q$$

During this pulsation, the matter can reach in its full extent neither the *Quantum System of Reference* as energy nor the system of reference of absolute rest as mass.

No event within the stationary system of absolute rest means *no time*. No frequency would be identified, since no-event (no-motion) could not result in photon impact.

At one end of its existence we might have the matter in its "absolute rest status". The absolute rest means the clear "mass status" of the matter, but the mass of the matter is only "measurable" in its relativistic meaning. We measure in fact the effects of the mass instead of real mass values. Therefore, we are not aware what the mass of the matter is like either in its "normal-measurable-relativistic" or its absolute rest status.

Ref
15A1
15A3

At the other end of its existence, we have the matter in motion, as energy quantum within the *Quantum System of Reference*. With reference to 15A1 and 15A3, the no-interaction and no-energy-exchange between photons means the frequency within this system of reference is equal to $f_q = 0$. The time count is indefinable. The *Quantum System of Reference* represents the clear energy status of the matter.

S
16.2.1

16.2.1 *The "speed" of the energy quantum*

Ref
S.14

With reference to Section 14, *photons* are the smallest energy quantum, results of the transformation of the mass of the matter into energy. Photons are the energy appearance of the matter.

Ref
S.14
S.15
Ref
15A1
15A3

Events happen in systems of reference. No event can however happen within a "system of reference" of a single quantum. With reference to Sections 14 and 15, photons, the smallest energy quantum, *are* impacted by events. They are the "messengers" of events. *Event means: interaction with photons.* With reference to 15A1 and 15A3, photons are not in impact or energy or work relations with each other. Single photons of equal energy quantum do not have their own time count. *Time means: number of photons of the impact.*

The energy (photons of equal energy quantum) is part of the matter, result of mass transformation. We may consider that all photons together compose a *Quantum System of Reference (QSR)*, with infinite number of single photons. Since all photons have the same velocity with no speed and no energy difference between them, we also may consider the *Quantum System of Reference* as subject for comparison with other systems of reference.

The comparison of any system of reference in motion with the *Quantum System of Reference* is unique and has two interpretations: The comparison of systems of reference to *QSR* or the comparison of the *QSR* itself to other systems of reference.

For practical reasons we use $\Delta v = a\Delta t = c - i$ for denoting the speed difference between the system of reference in motion with $i = \lim a\Delta t = c$ and the *Quantum System of Reference*. $(c - i)$ is an infinitely small and constant speed difference of the acceleration.

We also use $(c - v)$ in quantum entropy formulas for the characterization of the speed difference, result of the sphere symmetrical collapse.

Both $(c - i)$ and $(c - v)$ are real speed differences relative to c and always relate to the system of reference of the mass status of the matter in motion.

In spite of the examples above, any speed deduction from or summarization to the speed of photons gives identical results:

$$c + v = c \quad \text{and} \qquad \text{16E1}$$
$$c - v = c \qquad \text{16E2}$$

The collision of a system of reference of measured mass in motion with v with an *energy quantum* of speed c results in speed c of the *energy quantum*. Do 16E1 and 16E2 mean in this case the end of the mathematical correctness of summarizing speed values? No!

On the contrary, they are the drives for finding the correct explanation:

Photons are not accelerated up and are not slowed down as a result of collision with systems of reference of mass in motion (event). Systems of reference in collision with photons neither add nor take off energy from the photons. Photons are and remain in any circumstances of *equal* and *constant* energy quantum. *Photons are the energy themselves!*

> The collision of systems of reference in motion with photons can only be characterized by the intensity of the collision.

> The intensity of the collision is the impact of the motion: the number of the photons of the collision. No motion (no event) theoretically means no collision.

> The measurement of the intensity is the *frequency*, the number of the

impacted photons in collision for unit period of time. If the frequency before the collision was $f_q = 0$, afterwards it will be a certain value of f_v, depending on the energy intensity of the motion of the system of reference.

The collision with photons
- from in front to the direction of the motion increases the frequency;
- from behind decreases the frequency.

Here we have to note that the frequency cannot be negative. Its value can be less after the collision than before, but cannot be less than zero, the original frequency of the photons of the *Quantum System of Reference* without impact.

It means that any collision from behind of the motion is only possible if there is an impact already in place within the *QSR*. It also means that the sphere symmetrical expanding acceleration is the "genuine form" of the motion, the transformation of mass into energy. The sphere symmetrical collapse is only possible if the sphere symmetrical expanding acceleration, meeting photons always from in front, has already made its impact at the quantum system of reference.

S.
16.2.2   *16.2.2. Time count within the system of reference at absolute rest*

We are taking an identical event simultaneously happening in two systems of reference, different in their speed of motion. Since the event is one and the same, the absolute energy of its occurrence within the two systems of reference will induct the same n number of photons.

16F1
$$\Delta E_{SOR1} = \Delta E_{SOR2} = n \cdot q$$

Because of the speed difference, the frequency of the impact on the photons, the energy intensities of the identical event within the systems of reference, will be different:

16F2
$$\Delta e_{SOR1} = \frac{\Delta E_{SOR1}}{\Delta t_1} = \frac{n}{\Delta t_1} q = f_1 \cdot q \quad \text{and} \quad \Delta e_{SOR2} = \frac{\Delta E_{SOR2}}{\Delta t_2} = \frac{n}{\Delta t_2} q = f_2 \cdot q$$

16F3
$$e_{SOR1} \neq e_{SOR2} \quad \text{and} \quad f_1 \neq f_2$$

ΔE denotes absolute energies, Δe the energy intensities and Δt the durations of the event within the two systems of reference. n is the number of the photons, f is the frequency of the impact of the n photons.

With reference to the category of the event concentration, the energy balance of events happening in systems of reference distinct in motion is: $\dfrac{e_{SOR1}}{z_{SOR1}} = \dfrac{e_{SOR2}}{z_{SOR2}}$

The systems of reference of absolute rest and the *Quantum System of Reference* are similar and yet different. In one, no event (no motion) means no time flow (absolute rest), in the other, the no impact means the time flow is indefinable (*QSR*).

Ref
S.16.1
Ref
14I1
16B4

With reference to Section 16.1 on the inflection point of the sphere symmetrical collapse, to 14I1 on the quantum entropy and to 16B4 on the change of the quantum entropy, the matter exists in two forms between these two *ends*: mass and energy quantum.

Is the continuity of the matter in systems of reference guaranteed? In other words, do the mass and the photons of the matter, in these two forms of its existence fill the entire space available? The answer will come in the next section.

16.3.
The space coordinate

S. 16.3

The matter in transformation in time has two forms of its appearance: mass and energy quantum.

The energy balance of the matter is about
-   the energy of the mass (stationary within the system of reference of the measurement); and
-   the energy of the photons.

With reference to Section 13 on the sphere symmetrical expanding acceleration of the mass (of the matter) and Section 16 on its collapse, the energy intensities only give the absolute balance if they are corrected by the corresponding event concentration, here in 16G1, in the denominator:

Ref S.13

$$\frac{mc^2}{1} = \frac{m_v c^2}{\sqrt{1-\dfrac{v^2}{c^2}}} = \frac{m_u c^2}{\sqrt{1-\dfrac{u^2}{c^2}}} = \dots = \frac{m_i c^2}{\sqrt{1-\dfrac{i^2}{c^2}}}$$

16G1

where the index of the mass values denotes the speed of the system of reference of the mass.

A closer look on 16G1 proves that the balance relates not only to the mass, rather to the whole matter, addressing also the photons, the equal energy quantum, the results of the mass transformation into energy. With reference to Section 14 and 14A7:

Ref S.14 14A7

$$\frac{mc^2}{1} = \frac{mc^2\sqrt{1-v^2/c^2}}{\sqrt{1-v^2/c^2}} = \frac{mc^2\sqrt{1-u^2/c^2}}{\sqrt{1-u^2/c^2}} = \dots = \frac{mc^2\sqrt{1-i^2/c^2}}{\sqrt{1-i^2/c^2}} = mc^2$$

16G2

The absolute energy value relates to both forms of the matter.

Ref S.14

But where is the energy part? With reference to Section 14, as earlier proved, the energy does not leave the "adiabatic box" with only the mass within it. The photons, the product of the transformation of the matter, the result of the sphere symmetrical expanding acceleration of the mass shall also be within this "adiabatic box".

What are the coordinates of this "adiabatic box" of the matter?

Before we answer this question, we have to remember that the continuity of the time flow, with reference to Section 13, is guaranteed by the continuity of the transformation of the matter. With the growth of the actual speed, the time flow speeds up:

16G3
$$dt_v = \frac{dt_o}{\sqrt{1 - v^2/c^2}}$$

Ref
S.1
S.3.1
3C2
3C3
Ref
S.1
S.2
S.3

The sphere symmetrical expanding acceleration to speed v, can be assessed from the point of view of both ends: the stationary system of reference and the system of reference in motion with v. Since this is one and the same event (motion), the "space" coordinates are physically also one and the same, just assessed from different points of view, depending on the status (motion or stationary) of the system of reference of the examination.

If the transformation happens in *SORto* for $\Delta t_o = 1$, it happens from the point of view of *SORtv* for time period as in 16G3. With reference to 3B3 and 3C2, the length of the acceleration, measured from the point of view of the system of reference in motion, is:

16G4
16G5
$$dy_v = at_o \frac{dt_o}{\sqrt{1 - \frac{a^2 t_o^2}{c^2}}} \quad \text{and} \quad y_v = \frac{c^2}{a}\left(1 - \sqrt{1 - \frac{a^2 t_o^2}{c^2}}\right)$$

16G6
Ref
14A8

$$\text{or} \quad y = \frac{c^2}{a}\left(1 - \sqrt{1 - \frac{v^2}{c^2}}\right) \qquad \text{where } v = a\Delta t_o$$

With reference to 14A8, the absolute value of energy, necessary for the transformation of a stationary mass equal to $m = m_o$ into motion with v, is:

16G7
$$W = \Delta E = n \cdot q = mc^2 - m_v c^2 = mc^2\left(1 - \sqrt{1 - \frac{v^2}{c^2}}\right)$$

where in 16G4 and 16G5 at $\Delta t_o = 1$ is: $\dfrac{mc^2}{\Delta t_v} = \dfrac{mc^2}{\Delta t_o}\sqrt{1 - \dfrac{v^2}{c^2}} = m_v c^2$ and $\dfrac{mc^2}{z_o} = \dfrac{m_v c^2}{z_v}$

Ref
13C3

With reference to 13C3, the work intensity of the acceleration (the transformed internal energy of the mass) from the point of view of the system of reference of the mass, is:

16G8
$$w_v = m_v c^2 - m_v c^2 \sqrt{1 - \frac{v^2}{c^2}} = m_v c^2\left(1 - \sqrt{1 - \frac{v^2}{c^2}}\right) = \Delta e = \frac{n \cdot q}{\Delta t_v} = f_v \cdot q$$

16H1
16H2
$$y_v = \frac{f_v}{m_v a} q \quad \text{or} \quad y_v = \frac{n}{ma} q \qquad \left[\frac{n}{\Delta t_v} = \frac{n}{\Delta t_o}\sqrt{1 - \frac{v^2}{c^2}}\right] \text{ and } \left[m_v = m\sqrt{1 - \frac{v^2}{c^2}}\right]$$

With reference to Section 15, we have to remember that the *work intensity* in systems of reference is, in fact, the unified work (or energy) value for a unit period of time. Therefore, 16H1 in correct dimension it would be:

$$y_v = \frac{f_v}{m_v a} H$$

16H3

where H is the Planck constant of the system of reference of the mass acceleration, which has the dimension of *Joule · sec*.
16H3 can be written as:

$$y_v = \frac{f_v}{a \cdot m\sqrt{1 - (v^2/c^2)}} H$$

16H4

With reference to the reciprocal speed between systems of reference, should we continue to denote the result in 16H1 through y_v, but mark the length in 16H2 through y_o we come to:

$$\frac{dy_o}{dt_o} = \frac{dy_v}{dt_v} \quad \text{and} \quad dy_v = \frac{dy_o}{z_v} = \frac{dy_o}{\sqrt{1 - v^2/c^2}}$$

16H5
Ref
S.1
S.2

16H5 corresponds to the relation of the relativistic space coordinates deduced in Section 1 and 2.

The results in 16H1 and 16H2 do indeed represent the length of the event. The length is the quotient of $\Delta E = n \cdot q$, the transformed energy or $\Delta e = f_v \cdot q$, the energy intensity (in correct dimensions $\Delta e = f_v H$) and:

$F_o = ma$ or $F_v = m_v a$

        - the driving force of the transformation of the matter

16G6 is correct in its form for both systems of reference: the stationary and also the one in motion, since the actual speed is reciprocal. *c=const* and with reference to Section 5, on the definition of relativistic and absolute masses, the acceleration for both systems of reference is also equal.

Ref
S.5
Ref
S.14.2
16G6
16H1
16H2

With reference to Section 14.2, the formulas in 16G6, 16H1 and 16H2 are in line with the definition and meaning of the quantum entropy.

As 16G6 suggests the acceleration cannot be zero: $a \neq 0$, $\dfrac{dv}{dt} \neq 0$

16H6

The acceleration is part of the existence of the matter. There is no time without motion, without the transformation of the matter. 16H1 and 16H2 also prove the full mass cannot be transformed. During the accelerating expansion and collapse, the matter exists in two forms: mass and energy. Their existence in balance means motion and time count.

And now we have to come back to the question posed before this sub-section: what is the size, the space coordinate of the "adiabatic box" of the matter, where all components are in balance?

Ref.
S.2
S.3

With reference to Section 2 and 3, photons do not have vector components in "their motion". With reference to Section 14, the matter either exists as energy quantum with the smallest energy content without mass or has mass energy and a speed less than c. The overall balance between the mass, the energy, the photons, the motion, the time count and the space coordinates predict the continuity of the matter. It does not allow having space without time count, or time without space, or mass without energy (photons) and (photons) without mass.

Trying to picture any space configuration would mean speculation, but having the matter as infinite as it is, the mass-photon balance and the continuity of the time "flow" guarantee the continuity of the space fully "occupied" with energy (photons) and mass. There could not be "empty room" in the "adiabatic box" of the matter. With reference to 16H4, the boundaries of the existence of the matter (mass and energy in time) are infinite in space : $\lim y_{v \to c} = \infty$

16H7   and measured, depending on the motion from $\lim y_{v \to 0} = 0$ to $\lim y_{v \to c} = \infty$

S.
16.4
16.4
The quantum space or quantum membrane

Ref   With reference to Section 13.2, we have the matter with
13.2   -   mass in various stages of transformation and
        -    photons of equal energy quantum, result of the mass transformation, the
             Quantum System of Reference.
The matter is in continuous mass-energy balance. The sensitivity of this energy balance is equal to the value of a single energy quantum.

Any interference with the *Quantum System of Reference* within the adiabatic "box" of the matter distorts the genuine mass-energy balance of the matter.

Let us suppose we "hit" n number of photons of the *Quantum System of Reference*. The absolute work of the impact is:

16I1 $n \cdot q = \Delta E = W$

We suppose that the origin of the "hit", the impact of n number of photons, is an event (motion) within *SORti*. *SORti* is a system of reference, mass of m in motion with $i = \lim a\Delta t = c$, acceleration for infinite time. *SORti* is part of the mass energy transformation of the matter, in energy balance with the *Quantum System of Reference*.

The number and the absolute energy of all photons of the *QSR* in balance with *SORti* in acceleration is:

16I2 $N \cdot q = mc^2 \left(1 - \sqrt{1 - \frac{(c-i)^2}{c^2}} \right);$

The frequency of the balance is:

$$\frac{dN}{dt_i} q = \frac{dmc^2}{dt_i}\left(1 - \sqrt{1 - \frac{(c-i)^2}{c^2}}\right) = f_i \cdot q \qquad \text{16I3}$$

We suppose that the "hit" is equivalent to the work of mass m_v, speed of v within and relative to *SORti*. It means that *SORti*, in order to maintain the balance of its own internal energy, must keep, in addition to its own acceleration, a portion of its own mass m_v in motion against the *Quantum System of Reference*.

The acceleration of mass m_v against the *Quantum System of Reference* needs work:

$$W = m_v c^2 \left(1 - \sqrt{1 - \frac{(c-i)^2}{c^2}}\right) = n_v^v \cdot q \qquad \text{16J1}$$

The intensity of this work accelerating m_v and the frequency would be:

$$w_v = \frac{dm_v}{dt_i} c^2 \left(1 - \sqrt{1 - \frac{(c-i)^2}{c^2}}\right) = \frac{dn_v^v}{dt_i} q = f_v^v \cdot q \qquad \text{16J2}$$

as $\quad \dfrac{dm_v c^2}{dt_v z_v}\left(1 - \sqrt{1 - \dfrac{(c-i)^2}{c^2}}\right) = \dfrac{dm_v c^2}{dt_i z_i}\dfrac{\sqrt{1-(v^2/c^2)}}{\sqrt{1-(v^2/c^2)}}\left(1 - \sqrt{1 - \dfrac{(c-i)^2}{c^2}}\right) = \dfrac{dn_v^v}{dt_i z_i} q$

$$dt_i = 1 \text{ and } z_i = 1 \qquad f_v \le f_i \text{ function of the mass value.}$$

This is not, however, the case here: At the time of the impact, m_v has already been part of the acceleration of *SORti*, the motion with

$$i = \lim a\Delta t = c .$$

The effect of the "hit", the work of m_v is an additional impact to the total number of N photons of the *Quantum System of Reference*, already balancing the motion of *SORti*.

The absolute work formula of the hit is equal to 16J1. The measured mass is the same, the frequency however is determined by z, the event concentration of the impact. The work intensity formula of the impact in *SORti*, the system of reference in motion with $i = \lim a\Delta t = c$ is:

Ref
16J1

$$w = \frac{dm_v c^2}{dt_i z}\left(1 - \sqrt{1 - \frac{(c-i)^2}{c^2}}\right) = \frac{dn_i^v}{dt_i z} q = f_i^v \cdot q \; ; \qquad z = \sqrt{1 - \frac{v^2}{c^2}} \qquad \begin{array}{c}\text{16K1}\\\text{16K2}\end{array}$$

16K1 gives: $\qquad\qquad f_i^v = \dfrac{f_v^v}{z}$

In the 16K1 intensity formula the time component in fact depends on the speed of the impact: $dt_x = dt_i z$

We can write down 16K1 in form:

16K3
$$w = \frac{\frac{dm_v}{z}c^2}{dt_i}\left(1 - \sqrt{1 - \frac{(c-i)^2}{c^2}}\right) = \frac{dn_i^v}{dt_i z}q$$

16K3 means that the speed does not increase the relativistic (acting) mass value of the impact. Instead, as it is as expected, z, the event concentration does, since the energy of the motion of mass m_v with speed v is taken as external relative to its own mass energy.

We have the intensities of the two impacts, the motion of *SORti* and the motion of m_v within it, both refer to *SORti*.

The impact of *SORti*, in motion with $i = \lim a\Delta t = c$, to the *Quantum System of Reference* generates frequency:

16K4
$$\frac{dm - dm_v}{dt_i}c^2\left(1 - \sqrt{1 - \frac{(c-i)^2}{c^2}}\right) = \frac{dN_v}{dt_i}q = f_i^N \cdot q$$

The impact of the motion of m_v with speed v within *SORti,* generates frequency :

16K5
$$\frac{dm_v c^2}{dt_i\sqrt{1 - \frac{v^2}{c^2}}}\left(1 - \sqrt{1 - \frac{(c-i)^2}{c^2}}\right) = \frac{dn_i^v}{dt_i}q = f_i^v \cdot q$$

Ref
16K4   We can compare now the frequencies generated by the system of reference
16I3   without (16K4) and with (16I3) mass m_v in motion:

depending on the value of m_v $N_v \le N$ and $f_i^N \le f_i$

Ref
16K5   -   Should v, the speed of the impact be *zero*, the event concentration in
16I3   16K5 would be $z=1$, and there would not be a difference in the
16K4   frequency of the N photons balancing *SORti*. In this case the process is
           one and the same and the sum of 16K4 and 16K5 gives the frequency
           value in 16I3:

Ref
16K5
$$(m - m_v)c^2\left(1 - \sqrt{1 - \frac{(c-i)^2}{c^2}}\right) + \frac{m_v c^2}{\sqrt{1 - (v^2/c^2)}}\left(1 - \sqrt{1 - \frac{(c-i)^2}{c^2}}\right) = N \cdot q$$

-   Should however v be any value other than *zero*, the frequency of n_i^v number of photons out of the total N, balancing the motion of *SORti* will be the value as advised in 16K5.

(There are two processes and two impacts on the *Quantum System of Reference* in this case: the acceleration of *SORti* and the additional impact of the additional motion of m_v.)

The system is adiabatic and the original energy of *SORti*: $E = mc^2$

16.3.1. The response is the Quantum Membrane

The photons are of equal energy quantum. Any impact to the photons of the *Quantum System of Reference* cannot increase the energy of the photons "on average". Photons remain the value of the smallest energy quantum. The energy impact results in frequency increase, the increase of the number of impacted photons.

And so we have a *Quantum System of Reference* with photons of equal energy quantum, where a part may be impacted and the other may not. Any direct motion of photons from the spot of the impact in certain directions would mean energy difference between the photons. But, with reference to Sections 13, 14 and 15, there is no energy difference between the photons. They are of equal energy quantum. The photons are in motion and represent energy without mass. Photons do not carry the energy of impacts. They *represent* the impact within the *Quantum System of Reference*.

Ref.
S.13
S.14
S.15

If photons are of equal energy quantum and do not carry energy how can the *Quantum System of Reference* keep the balance and respond to any "external mass" impact?

The only answer is: *Quantum Membrane.*

Should the energy balance of the *Quantum System of Reference*, with sensitivity of a single energy quantum be impacted, the whole *QSR*, the "quantum space" as a *membrane* of photons of equal energy *quantum* will be responding. The response goes through the *Quantum System of Reference* as an acting *membrane* and photons will address the impact.

The *transfer* of the *energy impact* by the *Quantum Membrane* is immediate and the speed of the transfer corresponds to c, the velocity of photons. The transfer of the energy impact itself is not an event!

The transfer by photons does not have vector components and the *Quantum Membrane (QM)* acts in any direction of the space.

The energy is part of the matter and the *Quantum System of Reference* is representing this part. The other part is the mass. The transformation of the matter from mass into energy and from energy into mass establishes the time. Any external impact (other than the natural process of the mass-energy transformation) may distract the natural balance of the matter in transformation. The *energy-mass-time* balance of the matter cannot allow this to happen: the *Quantum Membrane* keeps and guaranties the immediate balance.

The generated frequency, the result of the impact, is part of the process of the natural *mass-energy-quantum-time balance* of the matter.

Once there is an impact within the *Quantum Membrane*, should energy be demanded, the *Quantum Membrane* will transfer it.

What is the distance of the response of the *Quantum Membrane*?

The measured distance of the *Quantum System of Reference* is relative. Therefore the size of the *adiabatic box* of the matter is relative. If there is no motion, there is no impact – the coordinates of the "space" are zero. The more is the speed of the motion, the longer are the measured "space coordinates" within (from the point of view of) the system of reference in motion.

The motion of the system of reference [of the mass (of the matter)] in collision with photons of equal energy quantum (the energy of the matter) determines the length of the "space coordinates".

The motion is the precondition of any space measurement. Without motion space coordinates have no meaning. The acceleration means motion and the motion obviously means time.

The dimensions of the space can be measured, but the space itself is not part of the matter. Since the dimensions of the space are determined by the motion and are relative, subject to measurement, the *size of the Quantum System of Reference* (the *Quantum Membrane*) varies from *zero* to *infinity*, depending on the motion of the system of reference of the measurement (which obviously must be different than the *Quantum System of Reference*).

The impact to the *Quantum Membrane* (the signal of the event) is not travelling within the "quantum space". It acts within the *Quantum System of Reference*. From the point of view of the system of reference of the impact, the time count has its specific meaning: the duration of the event. The event is the sequence of impacts caused to the *Quantum Membrane*.

What is in this case the length of an impact in space?

With reference to 14D8 and 16H4, the measured length within systems of reference in motion depends on the intensity of the work (or energy) of the impact.

16L1   The space coordinate is: $$y_v = \frac{w}{q \cdot a \cdot m \sqrt{1 - \left(v^2/c^2\right)}} H$$

The length of the transfer of the impact, the measured space coordinate, with reference to 16L1 corresponds to the intensity of the work and the acceleration of the system of reference. The impact is an energy, which can only be detected within length y_v and cannot be detected outside of this measured coordinate.

In collision with the quantum "particles" of the matter, the impacting system of reference in motion has its certain time systems (count). This system of reference is impacting a certain number of photons of the *QM* for the unit period of time, generating frequency. The impact of the collision will be transferred all over the *Quantum Membrane*.

If detected, the energy can be transferred into motion of mass again. If not, it stays as quantum energy within the *Quantum System of Reference*.

The distance of a space is a measured value, function of the motion. If no motion, no impact, no time, no space.

At the same value of acceleration the speed of the motion can vary. With reference to 16H7, the motion determines the measured size of the "adiabatic box" of the matter.

S.17   **17**

The process, the particles and the structure

The transformation of matter *starts* with sphere symmetrical expanding acceleration. With reference to Section 14, it produces photons of equal energy quantum. This phase lasts until the acceleration of the measured mass reaches $i = \lim a\Delta t = c$. The mass value, transformed into energy quantum, is:

17A1
$$\Delta m_{trans} = m\left(1 - \sqrt{1 - \frac{i^2}{c^2}}\right);$$ where m is the original mass of the matter at rest

The mass of the non-transferred matter at $\lim i = c$ is:

17A2
$$m_i = m\sqrt{1 - \frac{i^2}{c^2}}$$

The energies of the masses in 17A1 and 17A2 give the balance: $E = mc^2$

The value of m_i obviously depends on the original mass value of the matter: Should $m_1 > m_2$, the resulting relation of the mass values at $i = \lim a\Delta t = c$ will also be similar: $m_{i1} > m_{i2}$.

The sphere symmetrical expanding acceleration at $i = \lim a\Delta t = c$ *continues* until the mass of the matter reaches the value of the quantum entropy. For a supposed mass value of m_i, the value of the quantum entropy at the end of its sphere symmetrical expanding acceleration (for infinite time), is equal to:

17A3
$$m_{rem} = m_i\sqrt{1 - \frac{(c-i)^2}{c^2}} = m_{qe}$$ (with reference to 14E4, 14H1 and 14I1)

The work intensity of the accelerating mass in 17A3 keeps balance with the energy of the photons of the *Quantum System of Reference*:

17A4
$$\frac{n}{dt_i}q = \frac{dW}{dt_i} = dw_i;\quad \text{and}\quad df_i \cdot q = dm_i c^2\left(1 - \sqrt{1 - \frac{(c-i)^2}{c^2}}\right)$$

m denotes the original mass at its stationary status

The photons of the *Quantum System of Reference* collide with the remaining mass of the matter in acceleration for infinite time.

The energy intensity difference between the beginning and the end of the sphere symmetrical expanding acceleration is: $\Delta e = m_i c^2 - m_{qe} c^2$ 17A5

$$\frac{de}{dm} = c^2 \qquad \text{gives indeed:} \qquad \Delta e = m_i c^2 \left(1 - \sqrt{1 - \frac{(c-i)^2}{c^2}} \right)$$ 17A6

$(c - i) = a\Delta t$ denotes the acceleration from i to c without the need of the definition of the duration and the acceleration.

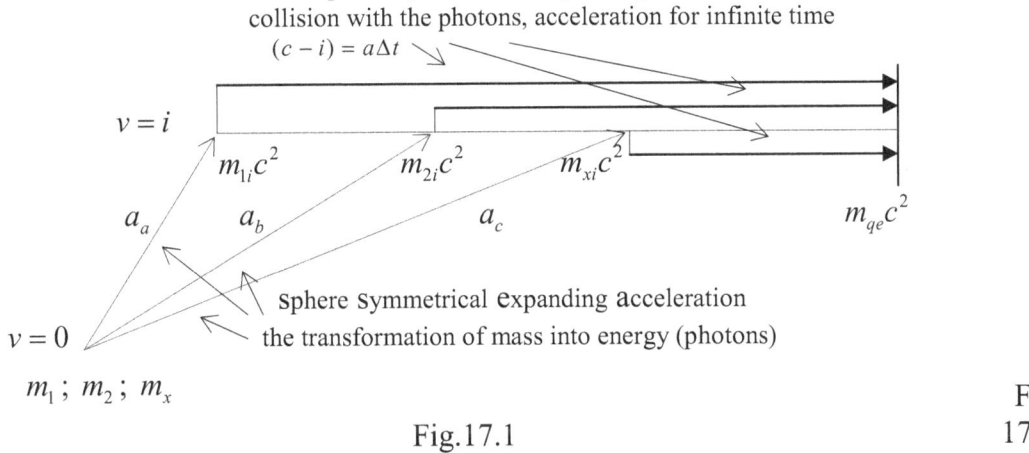

Sphere Symmetrical expanding acceleration
collision with the photons, acceleration for infinite time
$(c - i) = a\Delta t$

Fig.17.1

Fig
17.1

Fig 17.1 shows the two statuses of the sphere symmetrical expanding acceleration for mass values at rest of m_1 ; m_2 ; m_x .

There is an infinite number of varieties for the acceleration of the mass values. Fig.17.1 identifies them as a_a ; a_b ; a_c .

The energy balance of the matter for the first two stages of acceleration, the transformation and the acceleration for infinite time, is:

$$E_{ac} = mc^2 \left(1 - \sqrt{1 - \frac{i^2}{c^2}} \right) + m_i c^2 \left(1 - \sqrt{1 - \frac{(c-i)^2}{c^2}} \right) + m_i c^2 \sqrt{1 - \frac{(c-i)^2}{c^2}} = mc^2 \qquad \text{17B1}$$

or, in its other form:

$$E_{ac} = \Delta m_{trans} c^2 + n \cdot q + m_{qe} c^2 = mc^2 \qquad \text{17B2}$$

17B2 means the energy of the matter at the end of the sphere symmetrical expanding acceleration, the motion with $i = \lim a\Delta t = c$, is composed from:

➢ $m_{qe} c^2 = m_{rem} c^2$ the energy of the remaining mass of the matter (with reference to 17A3);

> $\Delta m_{trans}c^2 + n \cdot q$ the transformed energy and the energy of the impacted photons of equal energy quantum.

As expected, the energy of the impacted photon in collision is:

17B3
$$n \cdot q = mc^2 - (\Delta m_{trans}c^2 + m_{qe}c^2) = mc^2\sqrt{1 - \frac{i^2}{c^2}}\left(1 - \sqrt{1 - \frac{(c-i)^2}{c^2}}\right)$$

We can assess the energy balance in 17B1 for any intermediate K stage of the acceleration, the motion with $i = \lim a\Delta t = c$ for infinite time.

Sphere symmetrical expanding acceleration,
acceleration for infinite time

Fig.
17.2

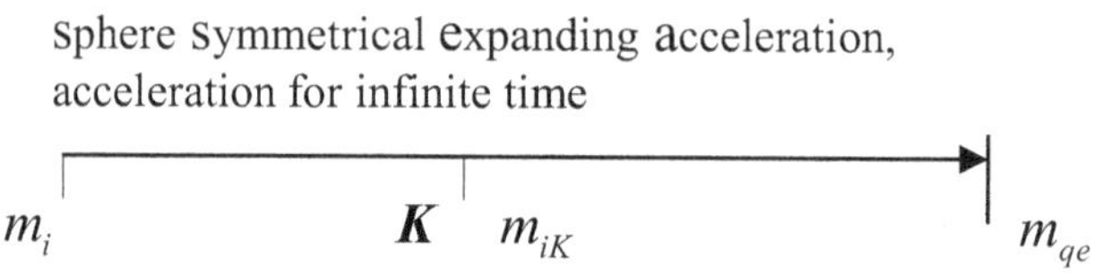

m_i $\boldsymbol{K}$ m_{iK} m_{qe}

Fig.17.2

With reference to 17A1, Δm_{trans} is the mass of the matter in transformation into energy quantum, reaching $\lim i = c$. Any mass value on the line between m_i and m_{qe} is the result of the acceleration of the mass of the matter for infinite time, and

17B4
$$\frac{de}{dm_i} = c^2; \quad \Delta e_K = m_i c^2\left(1 - \sqrt{1 - \frac{(c-i)^2}{c^2}}\right), \quad \text{where } m_{iK} = m_i\sqrt{1 - \frac{(c-i)^2}{c^2}}$$

Ref
S.14.2

The motion *is ruled* by the actual value of the acceleration. With reference to Section 14.2, it continues until $\dfrac{dF}{da} < m_i\sqrt{1 - \dfrac{(c-i)^2}{c^2}}$.

17B5
$$n \cdot q = mc^2 - (\Delta m_{trans}c^2 + m_{iK}c^2) = mc^2\sqrt{1 - \frac{i^2}{c^2}}\left(1 - \sqrt{1 - \frac{(c-i)^2}{c^2}}\right)$$

The number of the impacted photons depends on the change of the mass,

17B6
$$\frac{dn}{dm} = const; \quad \text{since} \quad \frac{c^2}{q}\sqrt{1 - \frac{i^2}{c^2}}\left(1 - \sqrt{1 - \frac{(c-i)^2}{c^2}}\right) = const$$

With reference to 17A4, the relation of the frequency and the change of the mass (energy) is constant:

17B7
$$\frac{dn}{dt_i} \cdot \frac{dt_i}{dm} = \frac{df_i}{dm_i} = const$$

As a natural feature of the existence of matter, the mass is being transformed into *energy*, into an infinite number of energy quantum of the smallest possible energy value with zero frequency: *The energy, alongside the mass, is part of the existence of matter.*

If we look at the proportions and at $E = mc^2$, the full *energy* of matter, the result of the transformation (the quantum energy) is the dominant part:

Quantum energy: Energy of the measured mass:

$$E_q = N \cdot q = mc^2 \left(1 - \sqrt{1 - \frac{i^2}{c^2}} \right) \qquad\qquad E_m = mc^2 \sqrt{1 - \frac{i^2}{c^2}}$$

17C1
17C2

> where N denotes the number of energy quantum, the result of the transformation of the mass of the matter into energy.

Ref
S.14.2

With reference to Section 14.2, the mass of the quantum entropy of a single photon is constant. With reference to

$$\frac{dF}{da} < m_i \sqrt{1 - \frac{(c-i)^2}{c^2}}$$

different summa mass values of the matter with the same acceleration result in different summarised mass values of quantum entropy. Equal summa mass values with different values of acceleration have the same summarised value of quantum entropy.

Once the energy of the mass in collision with photons is less than the energy of the quantum entropy, the energy of the photons prevails. The process steps into its *third stage.* It continues with the sphere symmetrical collapse of the remaining mass of the matter.

The energy impact of the *Quantum System of Reference*, the collision with n photons is external relative to the energy intensity of the collapsing mass (of the quantum entropy). The work of the collapse is:

$$\Delta E_{col} = \frac{m_i c^2 \sqrt{1 - \frac{(c-i)^2}{c^2}}}{\sqrt{1 - \frac{(c-v)^2}{c^2}}} - m_i c^2 \sqrt{1 - \frac{(c-i)^2}{c^2}}$$

17D1

where v is the absolute speed of the collapsing mass.

Ref
S.16.1
16B4

The collapsing mass at any $(c-v)$, the actual speed difference of the collapse relative to the QSR, with reference to 16B4, is:

$$m_{col} = \frac{m_i \sqrt{1 - \frac{(c-i)^2}{c^2}}}{\sqrt{1 - \frac{(c-v)^2}{c^2}}} = \frac{m \sqrt{1 - \frac{i^2}{c^2}}}{\sqrt{1 - \frac{i^2}{c^2}}} \sqrt{1 - \frac{(c-i)^2}{c^2}} = m \sqrt{1 - \frac{(c-i)^2}{c^2}} \;;$$

17D2

With reference to Section 16.1, v, the absolute speed in 17D2 cannot be zero $(v \neq 0)$, but as $\lim v = 0$ in this case $\quad c - v = i$

The energy balance of the matter in collapse is equal to

17E1
$$E_{col} = \Delta m_{trans}c^2 + n \cdot q - \Delta E_{col} + m_{col}c^2$$

➤ the difference of the energy of the *Quantum System of Reference* of the matter, the sum of the energies of the transformed and impacted in collision photons $(\Delta m_{trans}c^2 + n \cdot q)$ and the energy, used for making the collapse to happen (ΔE_{col}); and

➤ the energy of the mass at the end of the collapse $(m_{col}c^2)$.

Ref
17B2
17D1
17D2

With reference to 17B2, 17D1 and 17D2 it is equal to

$$E_{col} = mc^2\left(1 - \sqrt{1 - \frac{i^2}{c^2}}\right) + mc^2\sqrt{1 - \frac{i^2}{c^2}}\left(1 - \sqrt{1 - \frac{(c-i)^2}{c^2}}\right) -$$

17E2

$$-\left(\frac{m_ic^2\sqrt{1 - \frac{(c-i)^2}{c^2}}}{\sqrt{1 - \frac{(c-v)^2}{c^2}}} - m_ic^2\sqrt{1 - \frac{(c-i)^2}{c^2}}\right) + \frac{m_ic^2\sqrt{1 - \frac{(c-i)^2}{c^2}}}{\sqrt{1 - \frac{(c-v)^2}{c^2}}} = mc^2$$

The energy of the matter in transformation, in its accelerating and collapsing statuses is equal and constant:

17E3
$$E = E_{ac} = E_{col} = mc^2$$

We have noted before that at the end of the process there is a shift toward the quantum energy: As the expanding acceleration and accelerating collapse continues with a new cycle, as a natural form of the existence of the matter, the starting mass value of the matter in each consecutive cycle is less and less. The quantum energy is accumulating.

Mass value at the beginning of the cycle is: m

17E4   at the end of the cycle is: $m\sqrt{1 - \frac{(c-i)^2}{c^2}}$

Equivalent mass value of the energy quantum at the beginning is: 0

17E5   at the end of the cycle is: $m\left(1 - \sqrt{1 - \frac{(c-i)^2}{c^2}}\right)$

At the end of the n cycle:
Mass: Energy quantum:

17E6
17E7
$$m\left(\sqrt{1 - \frac{(c-i)^2}{c^2}}\right)^n \qquad m\left[1 - \left(\sqrt{1 - \frac{(c-i)^2}{c^2}}\right)^n\right]$$

The driving force of the transformation is:

$$F = \frac{dp}{dt}$$ where p is the momentum of the transformation-acceleration.   17F1

The energy balance for the count of the internal energy of the mass is:

$$\left(\frac{dm}{dt} v + \frac{dv}{dt} m \right) ds = dmc^2 \qquad\qquad ds = vdt$$ 17F2

where $v = at$ is the actual speed of the acceleration and s is the distance of the acceleration.

Making the necessary substitutions:

$$F = \frac{c^2 - v^2}{v} \frac{dm}{dt}$$ 17F3

17F3 means: The higher the intensity of the change of the mass, the higher the accelerating force is.

$$\frac{c^2 - v^2}{v} \frac{dm}{dt} = -am\left(1 - \sqrt{1 - \frac{v^2}{c^2}} \right);$$ where m is the original mass of the matter   17F4

$$\frac{dm}{dt} = -a \frac{m}{c^2 - v^2} v \left(1 - \sqrt{1 - \frac{v^2}{c^2}} \right); \qquad \text{and } \Delta m_v = m - m \cdot e^{-\left[\frac{v^2}{c^2 - v^2}\left(1 - \sqrt{1 - \frac{v^2}{c^2}} \right) \right]}$$ 17F5

Δm_v is the actual value of the transformed mass at reaching speed v of the acceleration.

17F5 means: for reaching the same value of speed, higher acceleration means higher intensity of the transformation of the mass into energy.

17.0.1. The three phases of the process   S
17.0.1

With reference to Section 13.2, the matter has an *infinite* number of *"mass-component-systems-of-reference"* in transformation. Should all mass components be in the same sequence, all systems of reference of the matter would accelerate and collapse in parallel. In this case all *mass-component-systems-of-reference* would reach the full transformation and full collapse at the same time.   Ref
S.13.2

Should however there be differences in the mass transformation, the values of the sphere symmetrical expanding acceleration and collapse there will be a shift between the cycles. This natural shift guarantees that the three statuses of the process are permanently present:

➤ *(a) the sphere symmetrical expanding acceleration of the mass of the matter*, the transformation of the mass into energy. This produces photons of equal energy quantum of zero frequency.
➤ *(b) the sphere symmetrical expanding acceleration of the mass for infinite time*, the motion with $i = \lim a\Delta t = c$. This is a balanced

energy status of the expanding mass and the *Quantum System of Reference*. The frequency of the impact is constant and corresponds to the motion with $\lim i = c$. No new photons are generated.

> ➤ *(c) the sphere symmetrical accelerating collapse of the mass of the matter*, under the effect of the *Quantum System of Reference*. The frequency of the photons at the collision grows.

Fig.17.3 shows the transformation of a single *mass-component-system-of-reference* of the matter.

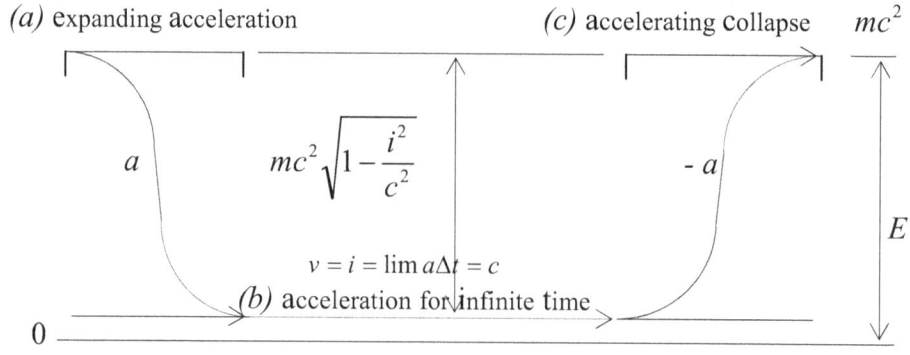

Fig
17.3

Fig.17.3

At the end of the sphere symmetrical expanding acceleration, at the inflection point of the acceleration, the mass components, equal in value to the quantum entropy are still in motion with $v=\lim i= c$. ($v < c$) The sphere symmetrical collapse (and slow-down, identified as negative acceleration) starts from this kinetic energy intensity status of matter. At the end of the slow-down (or acceleration relative to the quantum system of reference), at the inflection point, from where the expanding acceleration will start again, the speed of the collapse, is $v > 0$.

As the result of the internal energy of the mass, the sphere symmetrical expanding acceleration either starts, working against the "external" energy of the existing photons, formatting in this case a smooth transition from negative to positive acceleration, or in the case of the missing photons, *at the very beginning*, with a *"big bang"*, producing photons with an acceleration value of $a = \infty$.

17.0.2. The time-frame of the transformation

The *Quantum System of Reference* is unique and universal. The distinguishing $v = c$ speed separates it from any other system of reference in motion. In collision with other systems of reference, the frequency of the photons is the indicator of the speed of the given system of reference. The smaller the resulting frequency of the collision, the closer the speed of the system of reference is to the *Quantum System of Reference*.

We take three systems of reference:
SORto, the stationary system of reference, theoretically in absolute rest,

SORtpr the system of reference of the processes of the matter, and
SORtobs the system of reference of our observation (of the processes).

We suppose that all measurements are made within *SORtobs*, the system of reference of the observation in motion with $i = \lim a\Delta t = c$. We also suppose that the measured speed of the motion of the mass particles within *SORtobs* is $v = \lim i = c$.

There is no contradiction. *SORtpr* and *SORtobs*, the two systems of reference, are in motion with $v = i = \lim c$ relative to each other:
We can write

$$i + i = ii \qquad\qquad \text{17G1}$$

the components of 17G1 are equally $i = \lim c$ and $ii = \lim c$.

Consequently, the time relation of *SORtpr* to *SORtobs* is:

$$t_{pr} = \frac{t_{obs}}{\sqrt{1 - (i^2/c^2)}} \qquad\qquad \text{17G2}$$

where t_{pr} – is the duration of the event within the system of reference of the process (transformation, acceleration for infinite time, collapse); and t_{obs} – is the duration, measured within the system of reference of the observation.
We denote the time count within the stationary system of reference by t_o

The energy relations of the three systems of reference are determined by the time relations and are established through the values of the event concentration:

$$E = \frac{Mc^2}{z_o} = \frac{m_{obs}c^2}{z_{obs}} = \frac{m_{pr}c^2}{z_{pr}} \qquad\qquad \text{17G3}$$

where M, m_{obs}, m_{pr} are the values of the mass within the three systems of reference respectively. The values of the event concentration are the inverse values of the time relations for these systems and respectively are:

$$z_o = 1; \quad z_{obs} = z_o\sqrt{1 - \frac{i^2}{c^2}}; \quad z_{pr} = z_o \cdot z_{obs}\sqrt{1 - \frac{i^2}{c^2}} \qquad \text{It is taken as:} \quad z_i = \sqrt{1 - \frac{i^2}{c^2}} \qquad \text{17G4}$$

The energy intensity relations within the systems of reference:

$$E = \frac{e}{1} = \ldots = \frac{e_{obs}}{z_i} = \frac{e_{pr}}{z_i z_i} \qquad\qquad \text{17G5}$$

17G5 means:
The energy intensity of the process within *SORtobs*, the system of reference of the observation, is

$$e_{obs} = \frac{e_{pr}}{z_i}; \qquad \text{meaning:} \quad \frac{1}{z_i} \text{ times more than within } SORtpr. \qquad \text{17G6}$$

The durations of the process to be measured within *SORtobs* and *SORtpr* are equally infinitely long. This is why the different phases of the process can be observed from the system of reference of the observation as *quasi constant* and *stable statuses* of the matter.

When we measure (the effect of) the mass of a particle within the system of reference of the observation and its value is m_{obs},

then, with reference to 17G6, the "to-be-measured" mass value within the system of reference of the process, relative to the mass value, within the system of reference of the observation, would be infinitely small indeed:

17G7
$$m_{pr} = z_i m_{obs} = m_{obs} \sqrt{1 - \frac{i^2}{c^2}}$$

Should the process, happening within *SORtpr* be continuing to happen within *SORtobs*, the measured energy intensity would be

Ref
17G6
$$\frac{1}{z_i} \text{ "times more" as given in 17G6}$$

and the duration of the event, with reference to 17G2, z_i "times less":

17G8
$$t_{obs} = t_{pr} \sqrt{1 - \frac{i^2}{c^2}}$$

S.
17.1
17.1
The particles

The mass, the number and the acceleration of the particles, determines the (measurable) presence of the matter, the *element*.

The matter means the measured mass and the photons, as the result of the mass transformation in time. The "*element*" is identified by the mass, the number of the particles and by their relations to the *Quantum System of Reference*.

We can have elements with particles from 1 to a certain finite real number. (Category infinity in this case does not work. An infinite number of mass particles within an element would not give room for the energy transformation. An infinite number of mass particles would automatically mean the full matter without energy.)

The particles of the element in sphere symmetrical expanding acceleration, in transformation of their mass into energy, we call *protons* and denote them through p.

Protons, after reaching speed $i = \lim a\Delta t = c$ continue their sphere symmetrical expanding acceleration for infinite time. We call these particles *electrons* and denote them through e. *Electrons* lose in mass as a result of keeping the balance with the *Quantum System of Reference*, but do not generate photons.

Once the mass of the electrons of the element reaches the status of the quantum entropy, the sphere symmetrical collapse starts. The particles will collapse under the effect of the *Quantum System of Reference*. The particles in sphere symmetrical accelerating collapse we call *neutrons* and denote them through n.

Protons are energy providers, transforming their mass into quantum energy. *Electrons* are "fighters" and keep the balance within the element with the energy of the *Quantum System of Reference*. *Neutrons* are energy catchers. Their collapse re-transforms the quantum energy into mass again. The new cycle starts with less mass.

Protons and *electrons* in their acceleration use their *internal* "mass energy". *Neutrons* in their collapse use the *external* energy of the *Quantum System of Reference*.

Speaking about sphere symmetrical expanding acceleration and collapse, our conventional mind may suppose a "central" spot at rest or motion, around which other particles are in expansion or collapse. This view is understandable, but false: there are infinite numbers of spots, *all* in expanding acceleration or accelerating collapse. With reference to 16G6, the acceleration of the mass of the matter cannot be zero.

Ref
16G6

The definition of the particles is only acceptable if we add: This definition is only for distinguishing purposes on a certain level of our relativistic approach. Otherwise it might suggest we limit the existence of the matter. Protons, electrons, neutrons (and many others) cannot be final restrictive categories. We cannot find "*the particle*". Instead we find the effect of the mass and the *Quantum System of Reference* (of the matter) and call it proton, neutron, electron and others. We can count them, but this count is no more than the assessment of the effect – the mass energy balance of the matter (at the level of our measurement and capabilities).

Should there be any spot, central or other, or place at absolute rest within the matter, it would mean no time, no event and, consequently, no matter. Therefore, the matter must have an infinite number of systems of reference in motion with $i = \lim a\Delta t = c$, the acceleration for infinite time and in motion at the same time with $i = \lim a\Delta t = c$ relative to each other.

$$\lim i + \lim i + ... + \lim i = c$$

17H1

17H1 means that a slowing down or speeding up from $\lim ii = (\lim i) = c$ to $\lim i = (\lim ii) = c$ is fully acceptable and correct. It means two systems of reference, out of the infinite number, in relative motion to each other by i: two "levels" of our relativistic assessments.

S
17.1.1

17.1.1. *The mass of the particles*

The mass of the protons within the process *(p=)* varies

17I1
$$\text{between} \quad m \quad \text{and} \quad m\sqrt{1-\frac{i^2}{c^2}}$$

The mass of the electrons within the process *(e=)* varies

17I2
$$\text{between} \quad m\sqrt{1-\frac{i^2}{c^2}} \quad \text{and} \quad m\sqrt{1-\frac{i^2}{c^2}}\sqrt{1-\frac{(c-i)^2}{c^2}} \; ;$$

The total mass value of the neutrons *(n=)* is

17I3
$$\frac{m\sqrt{1-\frac{i^2}{c^2}}\sqrt{1-\frac{(c-i)^2}{c^2}}}{\sqrt{1-\frac{(c-v)^2}{c^2}}} = \frac{m\sqrt{1-\frac{i^2}{c^2}}\sqrt{1-\frac{(c-i)^2}{c^2}}}{\sqrt{1-\frac{u^2}{c^2}}}$$

Here in 17I3 we have to change the expression within the denominator.

17I4
$$u = c - v$$

There is no difference in the mathematical meaning, but the collapse of neutrons happens relative to the *Quantum System of Reference*. With reference to Section 17.0.1, speed difference *u* in accordance with our relativistic view does not correspond to absolute speed *v*. Therefore the change in the formula is certainly necessary.

The mass of neutrons varies between

17I5
$$m\sqrt{1-\frac{i^2}{c^2}}\sqrt{1-\frac{(c-i)^2}{c^2}} \; ; \quad \text{when } u = 0$$

the mass of neutrons equal to the mass of electrons (means: no collapse)

17I6
$$m\sqrt{1-\frac{(c-i)^2}{c^2}} \; ; \quad \text{when } u = i \text{ , it means the end of the neutron collapse.}$$

Ref
17E4
17E6

The mass value at the end of the collapse in 17I6 is in line with the mass values in 17E4 and 17E6.

There is *a difference* between the mass balance of the process and the summarized mass (or the effect of the mass values) of the element!

Since the particles are present in the transformation, acceleration and collapse in parallel, they all are "measured" together within the total mass of the element:

17I11
$$p \cdot k + e \cdot j + n \cdot r = M$$

where *M* is the atomic weight of the element, *k, j* and *r* denote the total number of protons, electrons and neutrons respectively within the element.

17.2.
Event concentration of elements, the rule for the change

S. 17.2

There are common milestones within the process. We take m for the mass of the protons at the beginning of the cycle

$$m \qquad m\sqrt{1-\frac{i^2}{c^2}} \qquad m\sqrt{1-\frac{i^2}{c^2}}\sqrt{1-\frac{(c-i)^2}{c^2}} \qquad m\sqrt{1-\frac{(c-i)^2}{c^2}} \ [= m_{(next)}] \qquad 17J1$$

proton   electron   neutron   [proton of the next cycle]

The change in the mass of the protons as a result of the transformation is

$$\Delta m_p = m - m\sqrt{1-\frac{i^2}{c^2}} = m\left(1-\sqrt{1-\frac{i^2}{c^2}}\right) \qquad 17J2$$

The change in the mass of the neutrons at $u = i$, at full collapse, is

$$\Delta m_n = m\sqrt{1-\frac{(c-i)^2}{c^2}}\left(\sqrt{1-\frac{i^2}{c^2}}-1\right) = -\Delta m_p\sqrt{1-\frac{(c-i)^2}{c^2}} \qquad 17J3$$

The change in the mass of the electrons is:

$$\Delta m_e = m\sqrt{1-\frac{i^2}{c^2}}\left(1-\sqrt{1-\frac{(c-i)^2}{c^2}}\right) \qquad 17J4$$

There is a mass-quantum energy balance in the electron process. But there is a mass difference between the proton expansion (transformation) and the neutron collapse (re-transformation). The difference in mass values suggests there is a difference between the intensities of the proton and the neutron processes.

The work intensity, necessary to accelerate mass Δm_p to $\lim i = c$ is:

$$w_p = \frac{\Delta m_p c^2}{\Delta t_p} = \frac{mc^2}{\Delta t_p}\left(1-\sqrt{1-\frac{i^2}{c^2}}\right); \qquad 17K1$$

or better to write: $\quad w_p = \dfrac{\Delta m_p c^2}{\Delta t_p} = \dot{m}_p c^2\left(1-\sqrt{1-\dfrac{i^2}{c^2}}\right) \qquad 17K2$

The source of the work intensity in 17K1 is the internal energy of the mass. It transforms into quantum status. It is measured within the system of reference of the proton, where t_p is the time measurement. (We measure intensities in systems of reference.) 17K2 is expressed in intensity terms. (From this point on, $\dot{m}$ will be used for intensity values.)

The work intensity of the neutron collapse, for slowing down Δm_n by

$\lim i = c$ relative to the *Quantum System of Reference* is

17K3
17K4
$$w_n = \frac{\Delta m_n c^2}{\Delta t_n} ; \quad \text{or} \quad w_n = \frac{\Delta m_n c^2}{\Delta t_n} = \dot{m}_n c^2 \sqrt{1 - \frac{(c-i)^2}{c^2}} \left(\sqrt{1 - \frac{i^2}{c^2}} - 1 \right)$$

The work intensity in 17K3 and 17K4 is measured within the neutron system of reference with time measurement t_n.

The slowing down by lim $i = c$ relative to the *Quantum System of Reference* during the neutron collapse gives exactly the speed difference of the acceleration from *zero* relative speed at the start of the proton acceleration up to lim $i = c$. We are comparing the proton-electron-neutron cycle. At the end of the neutron collapse the new cycle starts with less proton mass.

The proton and neutron processes must be in balance otherwise the continuity of the cycle cannot be guaranteed. (The end status of the neutron process and the start of the next proton cycle is the same.)

With reference to 17I1 and 17I3, the mass values of the proton and neutron processes are different. For ensuring the mass-energy balance of the matter, there must be intensity difference between the expanding acceleration of the protons and the accelerating collapse of the neutrons. The intensity difference also means difference in the duration of the two processes.

17L1   The intensity of the proton process: $a_p = \dfrac{i}{\Delta t_p} = \varepsilon_p$

17L2   The intensity of the neutron process: $a_n = \dfrac{i}{\Delta t_n} = \varepsilon_n$

We can compare work intensities, measured in different systems of reference if corrected by intensity values (or event concentration).

17L3
$$\frac{\dot{m}_p c^2}{\varepsilon_p} \left(1 - \sqrt{1 - \frac{i^2}{c^2}} \right) = \frac{\dot{m}_n c^2}{\varepsilon_n} \sqrt{1 - \frac{(c-i)^2}{c^2}} \left(\sqrt{1 - \frac{i^2}{c^2}} - 1 \right)$$

17L3 means the expanding acceleration and accelerating collapse are in balance with each other (and with the *Quantum System of Reference*).

It is important to note that intensities ε_p and ε_n in 17L3 relate to the entire process, to the full collapse of the neutron and the full transformation of the proton. With reference to Section 6, time and intensities (and event concentration) values are functions of the speed of the motion.

The intensity relation of the neutron and proton processes is:

17L4
$$\frac{\varepsilon_n}{\varepsilon_p} = -\frac{\dot{m}_n}{\dot{m}_p} \sqrt{1 - \frac{(c-i)^2}{c^2}}$$

Let us suppose we measure the work intensities of the proton and neutron processes in 17K2 and 17K4 within (or from) *SORt(obs)*, a system of reference of our observation. We suppose that *SORt(obs)* is stationary relative to the systems of reference of the particles. (It means that *SORt(obs)* may also be in motion with $i = \lim a\Delta t = c$, in acceleration for infinite time.) Since the systems of reference of the proton and neutron are in motion with $\lim i = c$ relative to *SORt(obs)* we shall adjust the measured data with the values of the event concentration.

We take it that the event concentration of the system of reference of the observation is $z_o = 1$. The event concentration of the systems of reference   17L5
in sphere symmetrical expanding acceleration and collapse, relative to *SORt(obs)* is:

$$z_i = \sqrt{1 - \left(i^2/c^2\right)}$$ 17L6

Because of the difference in event concentration, the measured value of the work intensity is z_i times more within the system of reference of the observation than within the system of reference of the particles. The duration of the event is also z_i times less.

The work intensity of the proton process within *SORt(obs)* is

$$P = \frac{\dot{m}_p c^2}{\sqrt{1 - \left(i^2/c^2\right)}}\left(1 - \sqrt{1 - \frac{i^2}{c^2}}\right)$$ 17M1

The work intensity of the neutron process within *SORt(obs)* is

$$N = \frac{\dot{m}_n c^2}{\sqrt{1 - \left(i^2/c^2\right)}}\sqrt{1 - \frac{(c-i)^2}{c^2}}\left(\sqrt{1 - \frac{i^2}{c^2}} - 1\right)$$ 17M2

17M1 and 17M2 are the measured work effects of the
- sphere symmetrical expanding acceleration of the protons; and
- sphere symmetrical collapse of the neutrons.

These measurements are the effect of these particles within the system of reference of the observation. *The effect of the mass (conditionally can be called as weight) is equal to the work, the result of the mass transformation (or re-transformation).*

It means, what we call proton and neutron (and as we will see in Section 18, electron as well) are in fact processes or events rather than static particles. This cannot be a surprise, since we always measure the effect of the mass, rather than the mass itself.

We measure the *event* (the effect) of the changing matter: the *transformation* from its mass status into energy (*proton process*) and its *re-transformation* from energy into mass (*neutron process*).

The relation between P and N is

17M3

$$\frac{N}{P} = \frac{\dot{m}_n}{\dot{m}_p} \sqrt{1 - \frac{(c-i)^2}{c^2}}$$

With reference to 17L4, the relation of the measured weight of the mass of the protons and neutrons in 17M3 is equal to the relation of the intensities of the two processes:

17M4

$$Z_{element} = \frac{N}{P} = \left| \frac{\varepsilon_n}{\varepsilon_p} \right|$$

Once we are aware of the weights (effect of the mass), we are also aware of the *intensities* of the proton and neutron processes. The quotient of the "weight values" gives the *event concentration* of the element. It shows the intensity of the neutron collapse relative to the proton transformation.

The less is the value of Z, the more is the energy reserve of the element. The more Z is, the less is the energy reserve, the more are the energy needs of the neutron collapse.

The values of the event concentration vary between elements. This is one of the most important characteristics, the indicator of the *energy* capability of the element.

A low value of event concentration means a long neutron process, with low intensity. It suggests that the photons of the proton transformation of the element are available for (re-)actions with other elements.

S.
17.3

17.3
The structure

The mass-energy transformation of the matter results in particles, following the sequence of:
-   transformation of mass into energy quantum (as proton stage);
-   acceleration for infinite time in balance with the quantum system of reference (as electron stage); and
-   sphere symmetrical collapse, the re-transformation of the energy into measurable mass (neutron stage).

Higher acceleration means higher intensity of the process at any stage.

Once the proton extension reaches $\lim i = c$, the *electron* stage, the energy intensity of the electron is not sufficient to keep itself within its earlier (original) proton position. The collision with photons of the *Quantum System of Reference* within the element pushes it at a distance, which corresponds to its balancing actual accelerating energy. The "pushing effect" of the photons of the collision depends on the sphere symmetrical expanding acceleration "energy" of the remaining mass value of the electron.

$$dm_e c^2 \left(1 - \sqrt{1 - \frac{(c-i)^2}{c^2}}\right) = f_i \cdot q = F_i \cdot l = \frac{ddm_e}{dt}(c-i) \cdot l + \frac{d(c-i)}{dt} dm_e \cdot l \qquad 17N1$$

where $f_i = (dn/dt)$ and, for simplicity, we take that $(dm_e/dt) = const$,

$$l = \frac{f_i}{dm_i \cdot a} q = \frac{c^2}{a}\left(1 - \sqrt{1 - \frac{(c-i)^2}{c^2}}\right) \qquad 17N2$$

where l – is the measured distance; m_e – is the measured mass of the electron; F_i – is the internal accelerating force of the electron; q – is the energy quantum and f_i – is its frequency.

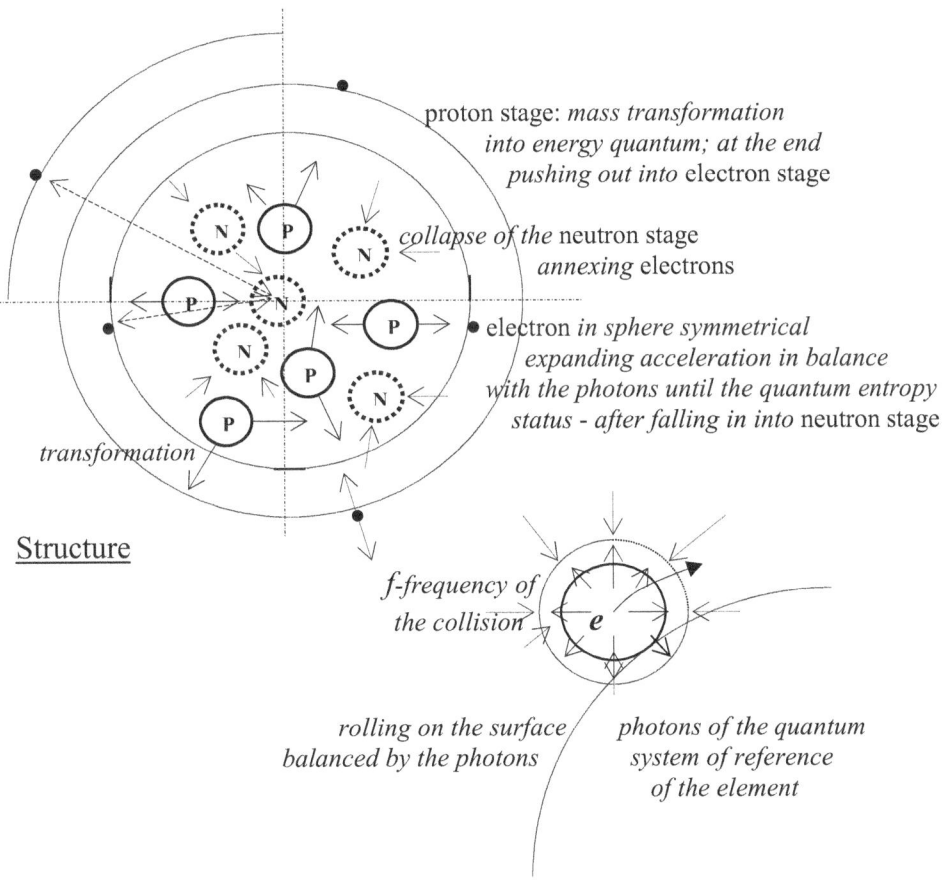

Fig.17.4

Fig. 17.4

In accordance with 17N3, the distance of the electron is inversely proportional to the value of its acceleration.

We have to note here that for low acceleration values, the distance in 17N2 approaches infinity. But 17N2 is calculated as measurable within the system of reference of the electron! Within a system of reference of the observation, relative to which the system of reference of the electron is in motion with lim $i = c$, this distance is within a range of real measurable dimension:

17N3
$$l_o = \frac{l}{z_i} = \frac{c^2}{a}\left(\frac{1}{\sqrt{1-\left(i^2/c^2\right)}} - 1\right)$$

The sphere symmetrical expanding acceleration of the electrons, in balance with the photons of the expanding protons rolls the electron around the accelerating protons and collapsing neutron of the element, we can call it *nucleus*.

Since the sphere symmetrical expanding acceleration is not perfectly symmetrical, the electron has a spin. Without the expanding acceleration of the electron, the angular momentum and the angular speed would be increasing.

In addition to the effect of the acceleration, we also have to note that from the side of the accelerating nucleus, one half of the semi sphere of the spinning electron moves "towards" the photon, resulting in an additional "blue shift" effect. The other half is spinning "away" from the photons resulting in the spin having an additional red shift effect.

As a result, the electron orbits around the nucleus, but this motion is the result of two effects: the protons of the nucleus in transformation and the spinning electron in sphere symmetrical expanding acceleration both roll the electron around the nucleus.

The electron finally reaches its quantum entropy stage, when it can no longer stand the "pressure" of the photons and the collapsing stage of the neutron starts. The sphere symmetrical expanding acceleration of the electron slows down and the electron "falls" into the nucleus as a neutron. In fact, the accelerating particles of the nucleus reach the electron, slowing down in its acceleration. This is no other than the annexation of the neutron by the expanding nucleus.

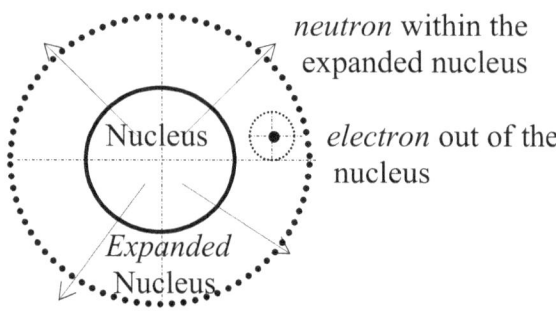

neutron within the expanded nucleus

electron out of the nucleus

Fig.
17.5

Fig.17.5

The nucleus contains accelerating protons and collapsing neutrons in a very unique way: The measurable mass of the protons is decreasing, the mass of the neutrons is increasing. Consequently, the expanding protons are *shrinking* and the collapsing neutrons are *expanding*. As the schematics in Fig.17.6 shows.

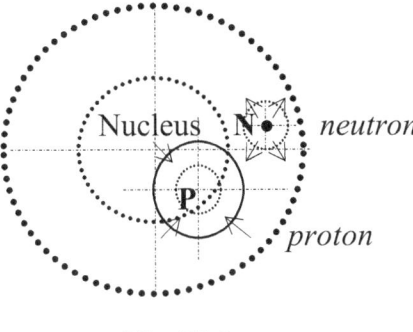

Fig.17.6

Fig
17.6

S.
18

18

Blue shift of the electron process

The mass-energy balance of the established proton-electron-neutron structure is the cohesive force of the atom of the element. The specifics of this energy balance determine the element. Matter is mass and energy, existing in time, in permanent generation, transformation and re-transformation into each other.

The proton process generates photons of equal energy quantum. The neutron process utilizes the quantum energy. The electron process is the collision of mass in sphere symmetrical expanding acceleration for infinite time with photons (of zero frequency). The result of the collision is *mass-energy* exchange: *blue shift* of photons at $i = \lim a\Delta t = c$. The work intensity of the blue shift is:

18A1
$$\frac{dn}{dt_i} q = \frac{dW}{dt_i} = w_i = \dot{m}_i c^2 \left(1 - \sqrt{1 - \frac{(c-i)^2}{c^2}} \right) \quad \text{and} \quad \frac{dm_i}{dt_i} = \dot{m}_i$$

18A2
$$f_i^{blue} \cdot q = \dot{m}_i c^2 \left(1 - \sqrt{1 - \frac{(c-i)^2}{c^2}} \right)$$

In 18A1 and 18A2 $\dot{m}_i$ is the intensity of the mass change of the electron.

Both sides of 18A1 and 18A2 are intensities and relate to the system of reference of the electron. The full expression is:

$$n \cdot q = mc^2 \sqrt{1 - \frac{i^2}{c^2}} \left(1 - \sqrt{1 - \frac{(c-i)^2}{c^2}} \right);$$

18A3
$$\frac{dn}{dt_i} q = f_i^{blue} \cdot q = \frac{dmc^2}{dt_i} \sqrt{1 - \frac{i^2}{c^2}} \left(1 - \sqrt{1 - \frac{(c-i)^2}{c^2}} \right) = \dot{m}_i c^2 \left(1 - \sqrt{1 - \frac{(c-i)^2}{c^2}} \right)$$

$$\text{where} \quad m_i = m\sqrt{1 - (i^2/c^2)} \quad \text{and} \quad dt_i = 1$$

The frequency is the intensity of the collision, the number of the collision of the mass with photons for a unit period of time.

The "blue shift" in 18A2 is not a real blue shift, since zero frequency cannot be blue shifted. It is in fact the result of the collision. We call it as blue shift for distinguishing purposes and for easy comparison.

Photons do not accumulate energy. They are, and remain, of equal energy quantum, the smallest possible. More frequency means more frequent collisions with the existing photons. With reference to Section 16, this non-accumulative feature of the *Quantum System of Reference* is the *Quantum Membrane*. The *QM* cannot withhold or conserve the energy impact and transforms it in the easiest possible way: the energy is directed to where the energy potential is the lowest. The lowest energy potential of the atomic structure is the neutron in its collapsing status.

The energy is provided to the neutrons by collision with photons in *red shift*. With reference to 15I5, 18A1, 18A2 and the conditions in 18A3:

$$\frac{dn}{dt_i}q = \frac{dW}{dt_i} = w_i = \dot{m}_i c^2 \sqrt{1 - \frac{(c-v)^2}{c^2}} \left(1 - \sqrt{1 - \frac{(c-i)^2}{c^2}}\right)$$

18A4
Ref

With reference to 18A2

18A2

$$f_v^{red} = f_i^{blue} \sqrt{1 - \frac{(c-v)^2}{c^2}}$$

18A5

The *red shift* is the process of taking energy by the neutrons from the *Quantum Membrane*. The consequence of the collision is the decrease of v, the absolute speed of the collapse. ($u = c - v$, the speed relative to the *Quantum System of Reference*, the result of the "negative" acceleration, as explained in the previous section, obviously increases.)

Ref
16D3

With reference to 16D3, the collision with photons during red shift results in the re-transformation of the quantum energy into mass. The collapse starts when the electron reaches the status of the quantum entropy.

18.1

Electron process

S.
18.1

The electron process of an element is a unified event. All atoms of the same element follow the same standards: the same number of electrons in sphere symmetrical expanding acceleration with the same intensity (for the same period of time).

$$m_i c^2 \left(1 - \sqrt{1 - \frac{(c-i)^2}{c^2}}\right) = n \cdot q$$

18B1

18B1 means: the mass change of the electron keeps balance with the energy of the photons.

We take it that this event happens for dt_s time. We also take it that this electron process corresponds to "normal" energy-mass balance conditions within the atomic structure. The intensity of the event is taken as $\varepsilon_s = 1$.

Normal energy-mass balance conditions means, the intensity of the electron process is in balance with the intensities of the proton and neutron processes. (Not "normal", unbalanced conditions are examined in Section 19.)

18B2
$$\frac{dm_i c^2}{dt_s \varepsilon_s}\left(1 - \sqrt{1 - \frac{(c-i)^2}{c^2}}\right) = \frac{dn}{dt_s \varepsilon_s} q = \frac{dn}{dt_i \varepsilon_i} q = n \cdot q$$

From 18B2 follows that for normal energy-mass balance case: $dt_s = dt_i$ dt_i corresponds to the speed of the sphere symmetrical expanding acceleration, the motion with $i = \lim a\Delta t = c$ and also $\varepsilon_i = \varepsilon_s = 1$.

Intensity ε_s in 18B2 and the intensities in 17L1, 17L2 and 17L3 are different by definition. The intensity in 18B2 characterises the electron process and is reciprocal to the time measurement, because the speed of the electron process is quasi constant.

For making easy comparisons between elements we take it that the mass change, relates to a single electron. Each element of the periodic table has its normal balanced electron process status. The mass of a single electron of each element is quasi constant:

18B3
$$m_s^e = \frac{e}{Periodic\ Number} = \frac{(M - P - N)_{mass-equivalent}}{Periodic\ Number} \approx const$$

The numerator in 18B3 is the total measured mass of the electrons of an element: the corresponding mass value of M, the atomic weight, P and N the measured effect of the proton and neutron processes (weights). The denominator is the periodic number.

The "atomic weights", the consequence of the mass change of the element, the protons and neutrons are measured within the system of reference of the observation. The result gives the mass of the electron, measured in this system of reference.

18B4
$$P = \frac{\dot{m}_p c^2}{\sqrt{1 - (i^2/c^2)}}\left(1 - \sqrt{1 - \frac{i^2}{c^2}}\right); \qquad N = \frac{\dot{m}_n c^2}{\sqrt{1 - (i^2/c^2)}}\sqrt{1 - \frac{(c-i)^2}{c^2}}\left(\sqrt{1 - \frac{i^2}{c^2}} - 1\right)$$

$\lim i = c$ cannot be derived from the measured mass values of the elements in 18B4. We are not aware of the intensity of the mass change within the systems of reference of the proton and the neutron.

S
18.1.1   *18.1.1. Intensity of the mass change of the electron process*

We are looking for the standard value of the intensity of the mass change of the electron in normal energy balanced circumstances. First, we have to return to Section 17.2 on the event concentration of elements.

Ref   With reference to 17M3 and 17M4
17M3
17M4
$$\frac{N}{P} = \frac{\dot{m}_n}{\dot{m}_p}\sqrt{1 - \frac{(c-i)^2}{c^2}}; \qquad and \qquad Z = \frac{N}{P} = \left|\frac{\varepsilon_n}{\varepsilon_p}\right|;$$

From 17M3 it follows that for any element:

18B5
18B6
$$Z \cdot \varepsilon_e = \sqrt{1 - \frac{(c-i)^2}{c^2}} = const \qquad and\ it\ is\ taken\ that \qquad \frac{1}{\varepsilon_e} = \frac{\dot{m}_n}{\dot{m}_p}$$

ε_e is the supposed intensity value of the electron process.

From 17L4 and 18B5 follows that

$$\left|\frac{\dot{m}_p c^2}{\varepsilon_p}\right| = \frac{\dot{m}_n c^2}{\varepsilon_n}\sqrt{1-\frac{(c-i)^2}{c^2}} \; ; \qquad \text{and} \qquad \frac{\dot{m}_p}{\dot{m}_n}\frac{\varepsilon_n}{\varepsilon_p} = \varepsilon_e Z \; ;$$

18C1
18C2

With reference to 18C1 and 18C2, the energy intensity difference between the mass change of the neutron and the proton can be written as:

$$\frac{\dot{m}_n c^2}{\varepsilon_n} - \frac{\dot{m}_n c^2}{\varepsilon_n}\sqrt{1-\frac{(c-i)^2}{c^2}} = \frac{\dot{m}_p c^2}{\varepsilon_e Z \varepsilon_p}\left(1-\sqrt{1-\frac{(c-i)^2}{c^2}}\right)$$

18C3

18C3 is the energy, equivalent to the mass change of the neutron, in balance with the electron process.

Ref
18B2

With reference to 18B2, the electron process is

$$\frac{dm_i c^2}{dt_s \varepsilon_s}\left(1-\sqrt{1-\frac{(c-i)^2}{c^2}}\right) = \frac{dm c^2}{dt_s \varepsilon_s}\sqrt{1-\frac{i^2}{c^2}}\left(1-\sqrt{1-\frac{(c-i)^2}{c^2}}\right) = \frac{dn}{dt_s \varepsilon_s}q$$

The electron process is of constant speed $i = \lim a\Delta t = c$, therefore denoting the mass change through dm is correct. It also can be written as

$$\frac{\dot{m}_e c^2}{\varepsilon_s}\left(1-\sqrt{1-\frac{(c-i)^2}{c^2}}\right) = \frac{dn}{dt_s \varepsilon_s}q$$

18C4

We cannot forget that for the electron process of each individual element, in "normal balanced" mass-energy conditions, the intensity is taken as $\varepsilon_s = 1$. Therefore

$$\dot{m}_e c^2\left(1-\sqrt{1-\frac{(c-i)^2}{c^2}}\right) = f_i \cdot q$$

18C5

With reference to 18C3 and 18C4, we establish the balance between the electron process and the energy intensity difference of the neutron and the proton mass change (in other words, the missing mass of the neutron):

$$\frac{\dot{m}_p c^2}{\varepsilon_e Z \varepsilon_p}\left(1-\sqrt{1-\frac{(c-i)^2}{c^2}}\right) = \frac{\dot{m}_e c^2}{\varepsilon_s}\left(1-\sqrt{1-\frac{(c-i)^2}{c^2}}\right)$$

18D1

18D1 can be written as:

$$\frac{dm c^2}{dt_p \varepsilon_n \varepsilon_e}\left(1-\sqrt{1-\frac{(c-i)^2}{c^2}}\right) = \frac{dm c^2}{dt_s \varepsilon_s}\left(1-\sqrt{1-\frac{(c-i)^2}{c^2}}\right)$$

18D2
18D3

$\varepsilon_e = \varepsilon_s$ obviously equal, since they relate to the same electron process.

18D3 brings us to the same statement in two different ways:
- with reference to 18B5

Ref
18B5

$$Z \cdot \varepsilon_s = \sqrt{1-\frac{(c-i)^2}{c^2}} = const$$

18D4

and since Z, the event concentration is different for each element, the *intensity* of the electron change is *different* for each element in "normal" energy balanced circumstances.

-    with reference to 18D2 and 18D3

18D5
$$dt_s = \frac{1}{dt_p \varepsilon_n}$$

and with reference to Section 17, the duration of the proton process and the intensity of the neutron process for each element are different. *It thus means the intensity of the electron process (the change of the mass of the electrons) for each element is different even in "normal" energy balanced circumstances.*

We take one of the elements for the basis of the comparison with others. The intensity of the electron process of this element, in normal energy balanced (standard) circumstances, is denoted by $\varepsilon_{se} = 1$. The event concentration of the electron process for other elements will be:

18E1
$$z_{sA} = \frac{\varepsilon_{sA}}{\varepsilon_{se}} ; \quad z_{sB} = \frac{\varepsilon_{sB}}{\varepsilon_{sB}} \dots \text{ and } \quad z_{sX} = \frac{\varepsilon_{sX}}{\varepsilon_{se}} \quad \text{where} \quad z_{se} = \frac{\varepsilon_{se}}{\varepsilon_{se}} = 1$$

where z_{se} is the event concentration of the electron process of the element taken for basis; z_{sA} to z_{sX} are the event concentration of the electron process of all other elements in normal circumstances.

The intensities of the electron process of elements are different, but the resulting frequency is always f_i, corresponding to the sphere symmetrical expanding acceleration, the motion with $i = \lim a\Delta t = c$. The event concentration of the electron process for each element is different:

18E2
$$z_{se} \neq z_{sA} \neq z_{sB} \neq \dots \neq z_{sX}$$

S.

18.2

18.2
Graphical interpretation of the proton-neutron-electron energy mass balance

Fig 18.1 shows a single cycle of the mass-energy transformation-re-transformation of the matter.

The mass statuses denote the milestones of the process in terms of stationary mass values. The effect of the intensity of each mass change corresponds to the measured weight of the particle of the process.

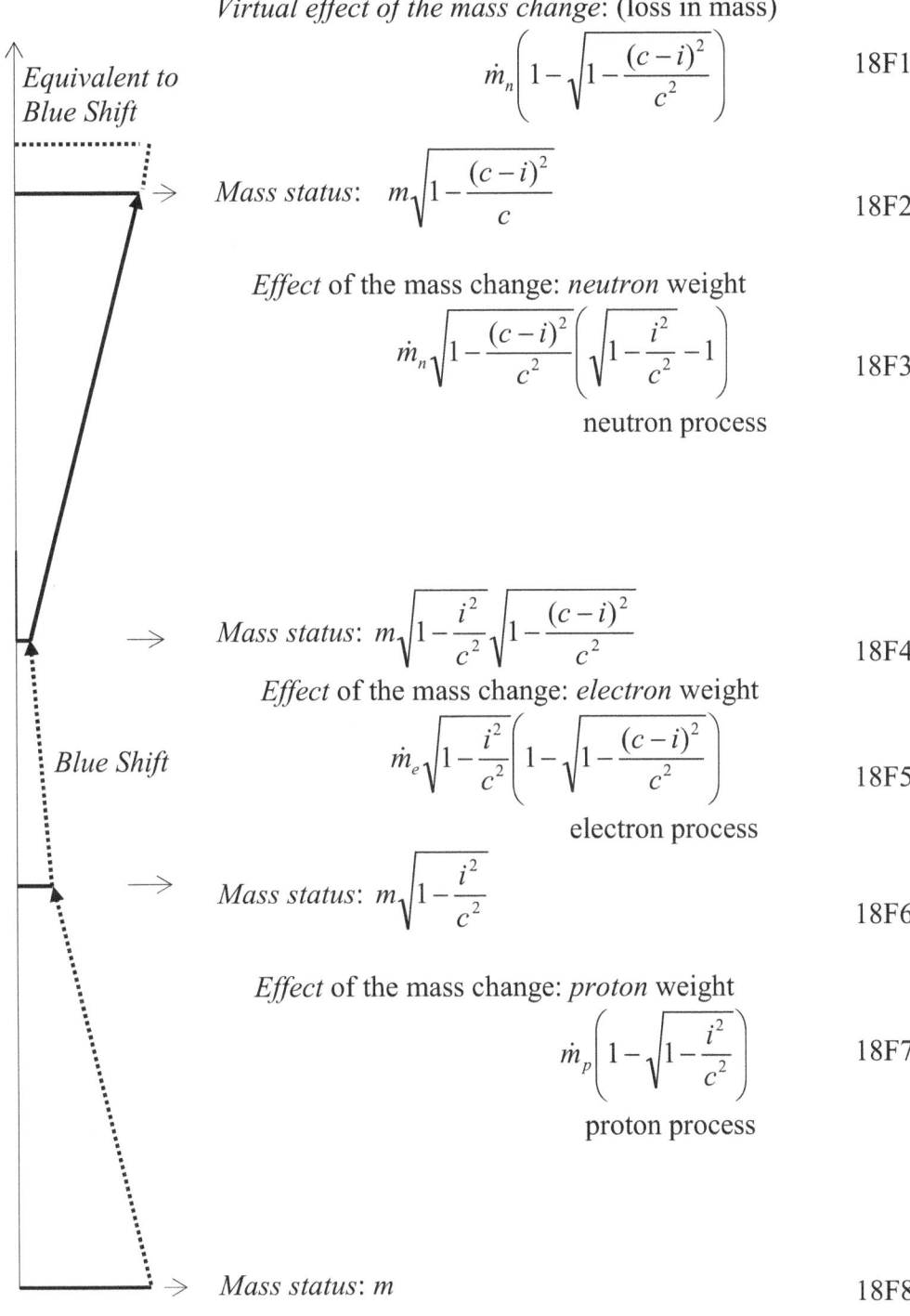

Virtual effect of the mass change: (loss in mass)

$$\dot{m}_n\left(1-\sqrt{1-\frac{(c-i)^2}{c^2}}\right)$$ 18F1

Equivalent to Blue Shift

Mass status: $m\sqrt{1-\frac{(c-i)^2}{c}}$ 18F2

Effect of the mass change: *neutron* weight

$$\dot{m}_n\sqrt{1-\frac{(c-i)^2}{c^2}}\left(\sqrt{1-\frac{i^2}{c^2}}-1\right)$$ 18F3

neutron process

Mass status: $m\sqrt{1-\frac{i^2}{c^2}}\sqrt{1-\frac{(c-i)^2}{c^2}}$ 18F4

Effect of the mass change: *electron* weight

$$\dot{m}_e\sqrt{1-\frac{i^2}{c^2}}\left(1-\sqrt{1-\frac{(c-i)^2}{c^2}}\right)$$ 18F5

electron process

Blue Shift

Mass status: $m\sqrt{1-\frac{i^2}{c^2}}$ 18F6

Effect of the mass change: *proton* weight

$$\dot{m}_p\left(1-\sqrt{1-\frac{i^2}{c^2}}\right)$$ 18F7

proton process

Mass status: m 18F8

Fig.18.1

Fig 18.1

There is an overlap in mass between the electron and neutron processes.

Ref 18F4   The sphere symmetrical expanding acceleration of the electron, in milestone 18F4, ends with mass, equal to *the mass of the quantum entropy*

$$m\sqrt{1-\frac{i^2}{c^2}}\sqrt{1-\frac{(c-i)^2}{c^2}}$$

and this is the mass value from where the neutron process starts.

The quantum entropy is the smallest energy represented by mass. Therefore, the mass in 18F4 may mean x number of particles, each of them with mass equal to the mass of the quantum entropy.

Since this overlap is repeating in each consecutive cycle, in cycle n the overlap in 18F4 will result in mass of quantum entropy:

18G1

$$m\sqrt{1-\frac{i^2}{c^2}}\left(\sqrt{1-\frac{(c-i)^2}{c^2}}\right)^n$$

The division of the total mass at the end of the electron process (at milestone 18F4) with the mass of the quantum entropy, would give the number of the neutrons. (In other words, the quotient of the mass at the end of the electron process and the number of neutrons is equal to the mass of the quantum entropy.)

There is a loss in mass (and energy) at the end of each cycle at milestone 18F1, equal to the effect (weight) of the intensity of the mass change of the neutron:

Ref 18F1

$$\dot{m}_n c^2\left(1-\sqrt{1-\frac{(c-i)^2}{c^2}}\right)=f^{blue}\cdot q$$

Ref 16D1 16D6 Ref 18C3 18D1   The energy of the mass in 18F1, in fact, has not been lost. It is equivalent to the blue shift of the photons, the result of the electron process. Without this blue shift the neutron collapse would stop and the process would become uncertain as presented in 16D1-16D6. With reference to 18C3 and 18D1, the blue shift is the external source, which provides the energy in the form of red shift to ensure the continuation of the neutron process.

18G2

$$\frac{\dot{m}_n c^2}{\varepsilon_n}\left(1-\sqrt{1-\frac{(c-i)^2}{c^2}}\right)=\frac{\dot{m}_e c^2}{\varepsilon_e}\sqrt{1-\frac{i^2}{c^2}}\left(1-\sqrt{1-\frac{(c-i)^2}{c^2}}\right)$$

The equation addresses the energy balance of the process: the virtual neutron mass change on the left hand side of 18G2.

The mass values of the process are equal at the milestones. Concerning the energy and mass balance of the process, however, we speak about the *effects* of mass changes rather than absolute mass values at rest. The intensities of the mass changes of the process give the measured "mass values".

Here we come to the proof of *Fermi's theory* of beta decay:

$$n \rightarrow p + e + \upsilon \qquad \text{or} \qquad n \rightarrow p + \beta + \upsilon \qquad \text{18G3}$$

The new cycle of the proton-electron-neutron process at milestone 18F2 starts with proton mass of

$$m\sqrt{1 - \frac{(c-i)^2}{c}} \qquad = (p) \qquad \text{18G4}$$

The effect of the missing neutron mass in 18G2 is equal to the electron process, meaning, to the measured electron mass:

$$\dot{m}_n c^2 \left(1 - \sqrt{1 - \frac{(c-i)^2}{c^2}} \right) \qquad = (e) \quad \text{or} \quad = (\beta) \qquad \text{18G5}$$

The equation in 18G2 refers to absolute values in order to address two events with different intensities. In 18G5 the intensity correction is taken off.

The absolute value of the mass overlap between the electron and the neutron processes, as presented in Fig.18.2 is

$$m \left(1 - \sqrt{1 - \frac{i^2}{c^2}} \right) \left(1 - \sqrt{1 - \frac{(c-i)^2}{c^2}} \right) \qquad \text{18G6}$$

18G6 in fact is a mass change, belonging to the proton and neutron processes. Its effect in intensity terms, with reference to 17L3, in balanced circumstances, gives zero:

$$\frac{\dot{m}_p c^2}{\varepsilon_p} \left(1 - \sqrt{1 - \frac{i^2}{c^2}} \right) - \frac{\dot{m}_n c^2}{\varepsilon_n} \sqrt{1 - \frac{(c-i)^2}{c^2}} \left(1 - \sqrt{1 - \frac{i^2}{c^2}} \right) = 0 \qquad \text{18G7}$$

In the case of any misbalance we can measure the a *neutrino* (υ), the effect of the mass unbalance.

As a conclusion we can establish that the neutron indeed transforms into a proton, as also presented in Section 17. The transformation may be accompanied by beta and neutrino decay.

A proton-electron-neutron cycle is shown in Fig.18.2. In order to prove the mass balance of the transformation, we use mass values in absolute terms. *m* is the mass of the proton at the start of the cycle.

18G5 and 18G7 well characterize the effect of the "measured particles". In the case of a missing electron of the mass balance, the *beta* decay is equivalent to the effect of a *positron*. (Since the mass change effect in 18G5 must be in balance with the electron process.)

18G7 addresses the cases of the missing mass with minus sign (*neutrino*) and with plus sign (*anti-neutrino*), which is equivalent to the *beta decay* of the *proton*.

Fig.18.2 shows the mass balance of the proton-electron-neutron process.

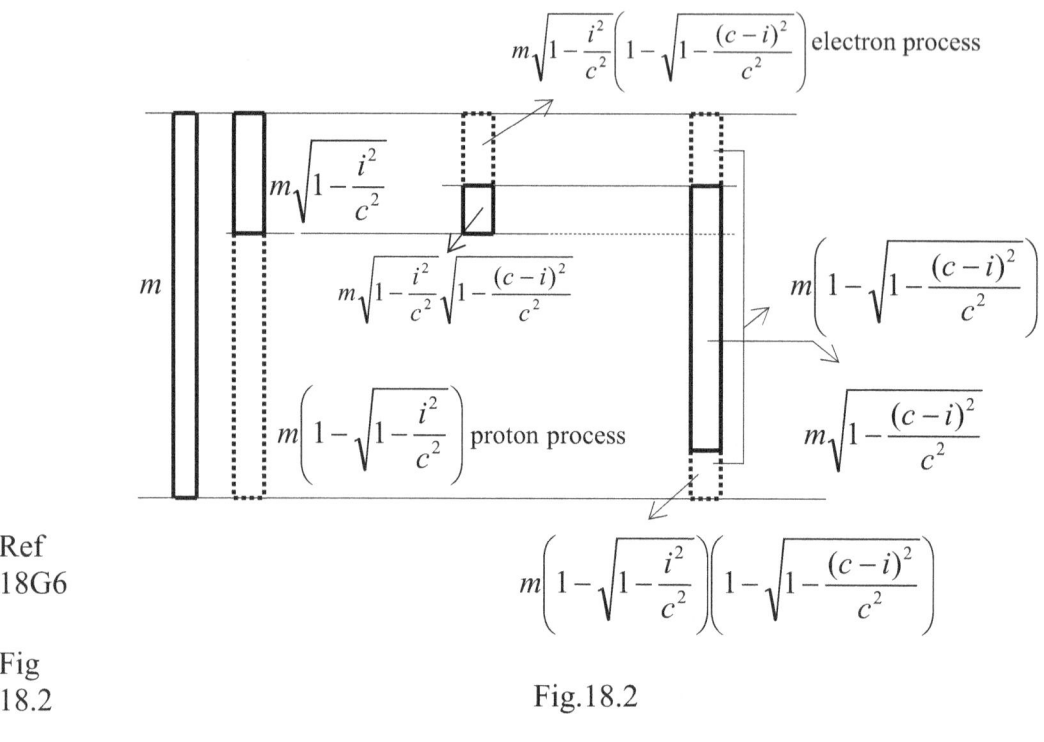

Fig.18.2

18.3.
Blue shift surplus and deficit of the electron process

What is the meaning of the frequency when divided by the intensity of the electron process?

$$\frac{dm_s^e c^2}{dt_X^e \varepsilon_X^e}\left(1-\sqrt{1-\frac{(c-i)^2}{c^2}}\right) = \frac{dn \cdot q}{dt_X^e \varepsilon_X^e}$$

where m_s^e - is the unified electron mass, dt_X^e - is the time of the electron mass change, ε_X^e - is the intensity of the electron process, n – is the number and q – is the energy of balancing energy quantum (photons).

The frequency itself characterises the intensity of the mass change. With reference to 18H1, in order to balance the mass change, n photons are necessary in the collision. The standard electron process is the sphere symmetrical expanding acceleration of the mass of the electron from $v = i$ to $v = c$ against the *Quantum Membrane*. The "energy" source of the process is the mass of the electron. This acceleration can happen in a less or in a more intensive way: With less or more value of acceleration.

Depending on the value of the acceleration this event need less or more time.

The resulting speed of the acceleration is constant. It is the motion with $i = \lim a\Delta t = c = const$. Therefore the effect of the mass change at the *Quantum Membrane* must also result in standard frequency, corresponding to $\lim i = c$ speed value. (The frequency relates to the time system, and the time system corresponds to the speed of the motion.) Thus, the time system of the acceleration, the motion with $\lim i = c$ for all electron processes is standard and equal.

The actual duration of the event (the acceleration of the mass) is a different category than the time system of the acceleration. The duration characterises the intensity of the acceleration.

$$E = \frac{\dot{m}_X^e c^2}{\varepsilon_X^e}\left(1 - \sqrt{1 - \frac{(c-i)^2}{c^2}}\right) = \frac{f_X}{\varepsilon_X^e}q \qquad \text{18H2}$$

When we divide the value of the mass change by the duration and get the mass change for the unit period of time (of the system of reference) – and call it *energy intensity*, as in 18H3

$$\dot{e} = \dot{m}_X^e c^2\left(1 - \sqrt{1 - \frac{(c-i)^2}{c^2}}\right) = f_X \cdot q \qquad \text{18H3}$$

this denotes the intensity of the event – as comparative data. The time system still relates to the sphere symmetrical expanding acceleration of the electron, the motion with $i = \lim a\Delta t = c = const$.

The energy *intensity* of the mass change is the real *intensity* of the process.

The *acceleration* value of the sphere symmetrical expanding acceleration is in fact the *intensity* of the motion with speed $i = \lim a\Delta t = c = const$ (the event). To compensate this intensity of the mass change $f_X = (n / dt_X^e)$ number of photons needed in collision for the unit period of time.

The higher intensity (higher acceleration value) means higher energy of the acceleration. And higher energy intensity needs more photons in collision. The time period of the mass change is reciprocal to the values of the acceleration (intensity), since the product of the acceleration and the time gives the speed difference of the acceleration, which is:

$$a \cdot \Delta t = (c - i) \quad \text{or} \quad \varepsilon_X^e \cdot dt_X^e = dv = (c - i) \qquad \text{18H4}$$

And this reciprocal character is standard for the electron process of all elements: $\qquad \varepsilon_X^e \cdot dt_X^e = ... = \varepsilon_Y^e \cdot dt_Y^e = dv = (c - i) \qquad \text{18H5}$

The *conflict* is that the frequency shall correspond to the time system of the system of reference in motion. If the motion is $i = \lim a\Delta t = c = const$, the time system and the frequency shall also be constant. With reference to the intensity of the event, the duration of the event is the one which generates different frequencies in the quantum collision.

The event needs the same number of photons, because the acceleration from $v = i$ to $v = c$ is standard and the mass is unified for all element. With low acceleration value the event takes longer, with high acceleration it takes shorter. If we divide the process with this duration we get the energy intensity of the event.

The intensity of the quantum response would be equal to f_X frequency.

The value of f_X however is *virtual*, since the quantum response cannot be higher (nor less) than the frequency of the corresponding speed of the motion, which is $i = \lim a\Delta t = c = const$.

Should we divide the energy intensity of the event on one side and the frequency of the quantum response on the other by the intensity of the process (reciprocal to the duration) we get back the values of the event in absolute terms.

➢ If the intensity of the electron process is equal to $\varepsilon_X^e = 1$, it gives exactly the frequency corresponding to $i = \lim a\Delta t = c = const$.

➢ If the intensity is more ($\varepsilon_X^e > 1$), there is a *surplus* in the electron blue shift (intensity) for use. (The virtual blue shift response would be more than the frequency corresponding to $i = \lim a\Delta t = c = const$.) It means in technical terms that the intensity of the neutron collapse of the element is less than the developing electron blue shift of the element. The element has blue shift *surplus*. It is detected though blue shift conflicts. (With reference to Section 20.1, classical elements are the *Hydrog*en and *Oxygen*.)

➢ If the intensity of the electron process is $\varepsilon_X^e < 1$, the element is in electron blue shift *deficit*. It does not mean at all that the existence of the element needs more blue shift. On the contrary: it is the distinguishing feature of the existence of the element. It means that the intensity of the neutron process is more than the intensity of the proton process. With reference to the Periodic table of elements and the Table 20.1 the majority of the elements have electron blue shift *deficit*.

The electron blue shift *surplus* and *deficit* are important characteristics of elements. They determine the activity of the element in external energy-mass balance relations (chemical reactions).

With reference to Section 19.1, the added proton mass to elements with electron blue shift deficit result in magnetic features. The added proton mass does not influence too much the elements with electron blue shift surplus, since those elements by their nature are in permanent blue shift conflict.

19

Magnetic structures

$\mathbf{T}$he *proton-electron-neutron process* is the cohesion force of elements.

The mass-energy (energy-mass) transformation (and re-transformation) is the existence of the matter itself. The matter has been in an infinite sequence of transformation cycles with different proton, neutron, electron masses and quantum energy. The effect and the relation of the mass and the energy (quantum) in time is the distinguishing characteristic of identified elements.

The mass of the transforming matter, with reference to Section 16 and 17E7, is less and less at the end of each cycle. The non-collapsing mass equivalent, the remaining quantum energy at the end of cycle η is:

Ref
S.16

$$mc^2\left[1-\left(\sqrt{1-\frac{(c-i)^2}{c^2}}\right)^{\eta}\right]$$

Ref
17E7

Each cycle starts with proton mass of:

$$m_{\eta} = m\left(\sqrt{1-\frac{(c-i)^2}{c^2}}\right)^{\eta-1}$$

19A1

where η denotes the number of the cycle and m is the proton mass

With reference to Section 18.1 and 18G2 the electron blue shift is equal to the "missing" mass of the neutron. The neutron collapse and blue shift balance, in cycle η, is:

Ref
S.18.1
18G2

$$\frac{\dot{m}_n c^2}{\varepsilon_n}\left(1-\sqrt{1-\frac{(c-i)^2}{c^2}}\right)^{\eta} = \frac{\dot{m}_e c^2}{\varepsilon_e}\sqrt{1-\frac{i^2}{c^2}}\left(1-\sqrt{1-\frac{(c-i)^2}{c^2}}\right)^{\eta}$$

19A2

here the intensities of the neutron and electron processes relate to cycle η

The steadily decreasing *mass* leads to increasing *quantum energy*. When the transformation cycle reaches the "last" value of quantum entropy, the collapse starts with steadily increasing *mass* and decreasing quantum *energy*.

Elements represent this infinite and permanent transformation/re-transformation process with particles, the measured effects of the proton, electron, neutron (and other) processes.

The time relation of the system of reference of the transformation and a supposed system of reference of the observation and measurement is:

19A3
$$t_{process} = \frac{t_{stationary}^{measurement}}{\sqrt{1 - \dfrac{i^2}{c^2}}} \; ; \quad \text{and} \quad t_{stationary}^{measurement} = t_{process}\sqrt{1 - \frac{i^2}{c^2}} \, ,$$

The *infinite* long process time within the system of reference of the particles corresponds to a certain measured lifetime (of the element or measured particles) and indicates an effect within the system of reference of the observation and measurement.

Ref
19A1

The effects of the proton, electron and neutron processes are the special characteristics of the elements. With reference to 19A1, the proton process starts with mass value, function of the number of the transformation cycle.

The infinite number of transformation cycles characterizes all statuses of the existence of the matter. At the same time we do not measure infinite number of elements.

We take Δm_p proton mass and make adjustments to the mass of a well known element. We are looking for the distinguishing effect of the adjustment.

S.
19.0.1

19.0.1. *The process with added proton mass*

Should *a proton mass be added* to an element, the element stays the same if the effects of the neutrons and the protons as well as the relation of the intensities of the neutron and proton processes stay constant.

To maintain the same element, with reference to 17M4, the event concentration must be equal to

Ref
17M4

$$Z = \frac{N}{P} = \left|\frac{\varepsilon_{n\psi}}{\varepsilon_{p\psi}}\right| = \left|\frac{\varepsilon_n}{\varepsilon_p}\right|$$

$\varepsilon_{n\psi}$ and $\varepsilon_{p\psi}$ are the intensities of the neutron and proton processes of the element; ψ denotes the processes with added proton mass.

With reference to Section 17K2, the effect of the normal proton process with added proton mass gives identical P "weight" values:

19B1
$$w_p = \dot{m}_p c^2\left(1 - \sqrt{1 - \frac{i^2}{c^2}}\right) = \dot{m}_{p\psi} c^2\left(1 - \sqrt{1 - \frac{i^2}{c^2}}\right) = w_{p\psi} = P$$

where the intensities of the proton transformation with normal and with additional proton masses are equal $\dot{m}_p = \dot{m}_{p\psi}$ 19B2

and value of:

$$\dot{m}_p = \frac{dm_p}{dt_p} \qquad \text{and} \qquad \dot{m}_{p\psi} = \frac{d(m_p + \Delta m_p)}{dt_{p\psi}}$$ 19B3
19B4

Δm_p is the added proton mass, dt_p and $dt_{p\psi}$ are the unit period of the mass change of the normal proton process and the proton process with added proton mass.

With reference to the equality of the effects of the measured mass in 19B1

$$\frac{dm_p}{dt_p} = \frac{d(m_p + \Delta m_p)}{dt_{p\psi}} \qquad \text{it follows:} \qquad dt_{p\psi} = \frac{d(m_p + \Delta m_p)}{dm_p} dt_p$$ 19B5

19B5 gives the time relation between the proton process with normal and added masses:

$$dt_{p\psi} > dt_p \text{ the process with added proton mass is longer.}$$ 19B6

Consequently the intensities are: $\varepsilon_{p\psi} < \varepsilon_p$,   19B7

the intensity of the proton transformation, with added mass is less.

Similarly to the proton mass change, the neutron process with added proton mass results in unchanged neutron process effect:

$$w_n = \dot{m}_n c^2 \sqrt{1 - \frac{(c-i)^2}{c^2}} \left(\sqrt{1 - \frac{i^2}{c^2}} - 1 \right) = \dot{m}_{n\psi} c^2 \sqrt{1 - \frac{(c-i)^2}{c^2}} \left(\sqrt{1 - \frac{i^2}{c^2}} - 1 \right)$$ 19C1

where the measured neutron mass with added proton mass is:

$$\dot{m}_{n\psi} = \frac{d(m_n + \Delta m_n)}{dt_{n\psi}} \qquad \text{and} \qquad \dot{m}_n = \dot{m}_{n\psi}$$ 19C2
19C3

$dt_{n\psi}$ is the unit period of the neutron process with added proton mass.

The larger mass acts for a longer period and with less intensity. It results in the same effect (which we may conditionally call "neutron weight") as the effect without added proton mass.

The proton-neutron energy mass balance for the element with added proton mass in absolute terms is:

$$\frac{d(m_p + \Delta m_p)c^2}{dt_{p\psi}\varepsilon_{p\psi}} \left(1 - \sqrt{1 - \frac{i^2}{c^2}} \right) = \frac{d(m_n + \Delta m_n)c^2}{dt_{n\psi}\varepsilon_{n\psi}} \sqrt{1 - \frac{(c-i)^2}{c^2}} \left(\sqrt{1 - \frac{i^2}{c^2}} - 1 \right)$$ 19D1

$\varepsilon_{n\psi}$ is the intensity of the neutron process with added proton mass

$\varepsilon_{n\psi} < \varepsilon_n$ the intensity of the neutron collapse with added proton is less than the intensity of the normal neutron process.

Ref
18F2
17L3
19D2

19D1 in other form, with reference to the neutron-proton process milestone in 18F2 and the mass balance in 17L3, is:

$$\frac{\dot{m}_{p\psi}c^2}{\varepsilon_{p\psi}}\left(1-\sqrt{1-\frac{i^2}{c^2}}\right)=\frac{\dot{m}_{n\psi}c^2}{\varepsilon_{n\psi}}\sqrt{1-\frac{(c-i)^2}{c^2}}\left(\sqrt{1-\frac{i^2}{c^2}}-1\right)$$

The absolute works of both sides of equations 19D1 and 19D2 are more than in normal (without added mass) circumstances. The stability of the element is, however, granted:

19D3

$$Z_{\psi}=\frac{\dot{m}_{n\psi}}{\dot{m}_{p\psi}}\sqrt{1-\frac{(c-i)^2}{c^2}}=\frac{N}{P}=\left|\frac{\varepsilon_{n\psi}}{\varepsilon_{p\psi}}\right|=\left|\frac{\varepsilon_{n}}{\varepsilon_{p}}\right|=Z$$

In spite of the added proton mass, the element remains the same.

It is easy to accept that in the case of a supposed *missing* proton mass the element would stay stable and unchanged in the same way. Analysing the electron-neutron process, however, we see that the impact of the adjusted proton mass is not without consequences.

Ref
18F6

The added proton mass results in a higher mass (number) of electrons, since with reference to the milestone in 18F6, the starting mass of the electron is equal to the final transformed mass of the proton:

19E1

$$m_{final\ proton}=m_{starting\ electron}=m\sqrt{1-\frac{i^2}{c^2}}\ ;\quad and\quad m_{e\psi}=(m_p+\Delta m_p)\sqrt{1-\frac{i^2}{c^2}}$$

The intensity of the electron process of the element with the added proton mass, however, remains the same (since this is still the same element):

19E2
Ref

$$\varepsilon_{e\psi}=\frac{\varepsilon_{p\psi}}{\varepsilon_{n\psi}}\sqrt{1-\frac{(c-i)^2}{c^2}}=\varepsilon_e$$

18F6
19B4

At the same time the mass of the electron, with reference to the proton-electron milestone in 18F6 with the added proton mass, is more.

Ref
18B2

With reference to 18B2, the blue shift of the unified electron mass for any element with normal energy mass balance structure is:

19E3

$$\frac{\dot{m}_ec^2}{\varepsilon_e}\left(1-\sqrt{1-\frac{(c-i)^2}{c^2}}\right)=\frac{dn}{dt_e\varepsilon_e}q=\frac{f_e}{\varepsilon_e}q=f_i\cdot q$$

since $dt_e=1$ and $\varepsilon_e=1$

$$dt_e\varepsilon_e=dt_s\varepsilon_s=...=dt_x\varepsilon_x\quad but\quad \dot{m}_ec^2\neq\dot{m}_sc^2\neq...\neq\dot{m}_xc^2$$

With reference to 18B3, the unified mass is taken as standard for each element in order to compare the intensities of the electron process independently of the mass of the electrons of the elements.

Ref
18B3

$$\frac{dm_s^ec^2}{dt_s\varepsilon_s}\left(1-\sqrt{1-\frac{(c-i)^2}{c^2}}\right)=\frac{dn}{dt_s\varepsilon_s}q=\frac{dm_e^ec^2}{dt_e\varepsilon_e}\left(1-\sqrt{1-\frac{(c-i)^2}{c^2}}\right)=\frac{dn}{dt_e\varepsilon_e}q=f_i\cdot q$$

In the case of an added proton mass to the element, the intensity of the electron process remains the same, but the electron mass is more and to balance it, it needs more photons in the blue shift balance:

$$\frac{\dot{m}_{e\psi}c^2}{\varepsilon_{e\psi}}\left(1-\sqrt{1-\frac{(c-i)^2}{c^2}}\right)=\frac{d(n+\Delta n)}{dt_{e\psi}\varepsilon_e}q \qquad \text{19E4}$$

There is a conflict in 19E4: the frequency of the blue shift at the speed of the sphere symmetrical expanding acceleration $i = \lim a\Delta t = c$ of electrons cannot be more than f_i. We continue with the effect of the blue shift and will return to this conflict later.

For the neutron collapse the energy is provided by the blue shift of electrons. The general equation is:

$$dmc^2\sqrt{1-\frac{i^2}{c^2}}\left(1-\sqrt{1-\frac{(c-i)^2}{c^2}}\right)=\frac{dmc^2\sqrt{1-\frac{i^2}{c^2}}\sqrt{1-\frac{(c-i)^2}{c^2}}}{\sqrt{1-\frac{(c-v)^2}{c^2}}}-dmc^2\sqrt{1-\frac{i^2}{c^2}}\sqrt{1-\frac{(c-i)^2}{c^2}} \qquad \text{19F1}$$

In 19F1, the left hand side of the equation is the blue shift, the right hand side is the work of the neutron collapse, provided by an external source (electron blue shift) source.

$$dmc^2\sqrt{1-\frac{(c-v)^2}{c^2}}\sqrt{1-\frac{i^2}{c^2}}\left(1-\sqrt{1-\frac{(c-i)^2}{c^2}}\right)=dmc^2\sqrt{1-\frac{i^2}{c^2}}\sqrt{1-\frac{(c-i)^2}{c^2}}\left(1-\sqrt{1-\frac{(c-v)^2}{c^2}}\right) \qquad \text{19F2}$$

Expressed differently, 19F2 gives the internal energy balance relation: the left hand side is the red shift of the neutron collapse and the right hand side gives indeed the internal work of the neutron collapse provided by the red shift of the neutrons.
In 19F1 and 19F2 m denotes the mass of the protons.

With reference to 16B1, the work of the sphere symmetrical collapse of a neutron within the element, in absolute terms, is:

$$W_{n\psi}=\frac{\dot{m}_{n\psi}c^2}{\varepsilon_{n\psi}}\left(\frac{1}{\sqrt{1-\frac{(c-v)^2}{c^2}}}-1\right)=\frac{dn_{n\psi}}{dt_{n\psi}\varepsilon_{n\psi}}q=n_{n\psi}\cdot q \qquad \text{19F3}$$

19F3 corresponds to the right side of the equation in 19F1. Its message is: the work for collapsing a neutron from electron stage with sphere symmetrical expanding acceleration at speed $i = \lim a\Delta t = c$ to speed v, is equal to the "energy" of $n_{n\psi}$ photons.

The neutron collapse with added proton mass to the process of the element needs work intensity of:

$$w_{n\psi}=\frac{\dot{m}_{n\psi}c^2}{\sqrt{1-\frac{(c-v)^2}{c^2}}}-\dot{m}_{n\psi}c^2 \qquad \text{19F4}$$

The work intensity in 19F4 is equal to the intensity of the normal neutron collapse, since with reference to 19C3 the neutron mass change intensities are equal – this is the same element:

Ref 19C3

$$\dot{m}_n = \dot{m}_{n\psi}$$

The difference between the normal and the process in 19F4 is the duration of the event. The neutron collapse with the added proton mass lasts longer, since with reference to 19D1: $\varepsilon_{n\psi} < \varepsilon_n$.

Ref 19D1

The red shifts for both processes therefore, in *intensity* terms, are equal:

19F5

$$w_{n\psi} = w_n = \frac{\dot{m}_n c^2}{\sqrt{1-\frac{(c-i)^2}{c^2}}} - \dot{m}_n c^2$$

But the *absolute work values* of the collapse are obviously different:

19F6

$$W_{n\psi} = \frac{\dot{m}_{n\psi} c^2}{\varepsilon_{n\psi}}\left(\frac{1}{\sqrt{1-\frac{(c-v)^2}{c^2}}}-1\right) = n_{n\psi}\cdot q > n_n \cdot q = \frac{\dot{m}_n c^2}{\varepsilon_n}\left(\frac{1}{\sqrt{1-\frac{(c-v)^2}{c^2}}}-1\right) = W_n$$

19F7   The *work intensities* are equal: $n_{n\psi}\cdot\varepsilon_{n\psi} = n_n \cdot \varepsilon_n$ and

$$\text{since}\quad \frac{c-v}{dt_{n\psi}} = \varepsilon_{n\psi}\quad\text{and}\quad\frac{c-v}{dt_n}=\varepsilon_n$$

The red shift frequencies for both cases are *equal*: equal number of photons, in collision for a unit period of time for both cases. (In absolute terms: different number of photons in collision, but acting for shorter and longer duration.)

With reference to 19F5, the element does not need higher frequency or more neutron work intensity than the normal in the case of added proton mass. The drive of the process is the blue shift of the electrons. And here we come back to the conflicting impact of the added proton mass on the blue shift of the electron process. In normal circumstances of the element, the intensity of the blue shift is:

19G1

$$w_e = \dot{m}_e c^2\left(1-\sqrt{1-\frac{(c-i)^2}{c^2}}\right) = \frac{dn}{dt_e}q = f_i \cdot q$$

Ref 19E4

In the modified circumstances, with reference to 19E4, the work intensity of the blue shift drive is:

19G2

$$w_{e\psi} = \dot{m}_{e\psi} c^2\left(1-\sqrt{1-\frac{(c-i)^2}{c^2}}\right) = \frac{d(n+\Delta n)}{dt_{e\psi}}q$$

19G3 $w_{e\psi} > w_e$

The work intensity of the electron process with the added proton mass is clearly more. The intensity of the demand of the neutron collapse remains the same, but will have a longer effect. The increased intensity of the electron process may destroy the element.

The conflict is that the required balancing frequency compensating the increased intensity of the electron process cannot result in frequency higher than f_i, the value, corresponding to $i = \lim a \Delta t = c$, the speed of the acceleration of the electrons.

With reference to 19G1 and 19G2, if

$$w_e = \frac{dn}{dt_e} q = f_i \cdot q \quad \text{the prediction is} \quad w_{e\psi} = \frac{d(n + \Delta n)}{dt_{e\psi}} q > f_i \cdot q \qquad \text{19G4}$$

which is impossible. The frequency cannot be more than the one corresponding to $i = \lim a \Delta t = c$. It would mean a lesser speed of the sphere symmetrical expanding acceleration, the motion with $\lim i = c$ for infinite time.

The effect of the increased intensity of the electron process is released permanently. The impact operates through the *Quantum Membrane*. The electrons impact the QM with f_i, a certain number of collisions for a unit period of time, and the *Quantum Membrane* cannot withstand the impact.

The energy balance conflict, however, is resolved.

S.

19.0.2. *The case with reduced proton mass* 19.0.2

The whole deduction can be repeated, for a case with a reduced by Δm_p proton mass. Following the same steps and logic, it is easy to accept that the event concentration of the element remains the same and the intensity of the electron process is equal to the normal process.

$$\varepsilon_{e\phi} = \frac{\varepsilon_{p\phi}}{\varepsilon_{n\phi}} \sqrt{1 - \frac{(c-i)^2}{c^2}} = \varepsilon_e \qquad \text{19H1}$$

Index ϕ denotes the case with missing Δm_p proton mass.

Obviously: $dt_e = dt_{e\psi} = dt_{e\phi}$

As in the previous cases, the proton and neutron processes are of the same work intensity. The effects of the proton and neutron changes result in the same effect as for the standard processes: the acting proton and neutron masses are less, but the intensities of the mass change are more.

$$w_{n\phi} = w_n = w_{n\psi} \quad \text{and} \quad w_{p\phi} = w_p = w_{p\psi} \qquad \begin{array}{c}\text{19H2}\\\text{19H3}\end{array}$$

But the intensity of the electron process is less than the red shift intensity demand of the neutron collapse. The frequency cannot be less than f_i the one corresponding to $i = \lim a \Delta t = c$:

$$w_{e\phi} = \dot{m}_{e\phi} c^2 \left(1 - \sqrt{1 - \frac{(c-i)^2}{c^2}} \right) = \frac{d(n - \Delta n)}{dt_{e\phi}} q \qquad \text{19H4}$$

Ref   With reference to 19E1
19E1
19H5
$$m_{e\phi} = (m_p - \Delta m_p)\sqrt{1 - \frac{i^2}{c^2}} \quad \text{and} \quad \dot{m}_{e\phi} < \dot{m}_e$$

The missing demand in the "added proton mass case" is provided by the "missing proton mass case". The missing blue shift in the second is provided by the first. The energy-mass balance is assured.

$$\frac{d(m_p + \Delta m_p)}{dt_{e\psi}\varepsilon_{e\psi}}\sqrt{1 - \frac{i^2}{c^2}}\left(1 - \sqrt{1 - \frac{(c-i)^2}{c^2}}\right) +$$

19H6
$$+ \frac{d(m_p - \Delta m_p)}{dt_{e\phi}\varepsilon_{e\phi}}\sqrt{1 - \frac{i^2}{c^2}}\left(1 - \sqrt{1 - \frac{(c-i)^2}{c^2}}\right) =$$

$$= \frac{d(n + \Delta n)}{dt_{e\psi}\varepsilon_{e\psi}}q + \frac{d(n - \Delta n)}{dt_{e\phi}\varepsilon_{e\phi}}$$

$$m_e = m_p\sqrt{1 - i^2/c^2}\ ; \ \varepsilon_e = \varepsilon_{e\psi} = \varepsilon_{e\phi}\ ; \ dt_e = dt_{e\psi} = dt_{e\phi}\ ; \text{ and } \ f_i = \frac{dn}{dt_e}$$

And, finally, for the complete case the electron blue shift balance is assured.

19H7
$$\dot{m}_e c^2\left(1 - \sqrt{1 - \frac{(c-i)^2}{c^2}}\right) = f_i \cdot q$$

The *Quantum Membrane* keeps the element in balance.

S.
19.0.3   *19.0.3. The balance of the process*

With added proton mass, the element has electron dominance: it means the same number of electrons – the characteristic of the element – but with higher intensity of the electron process, higher effect (more "weight").

With missing proton mass there is an electron shortage in the process: the same number of electrons, but with less intensity of the process. The effect is weaker and the virtual weight of the electrons is less.

The electrons are in acting mass-energy balance with the *Quantum Membrane*. Electron dominance provides blue shift surplus to the *Quantum Membrane*. The case with electron shortfall works with blue shift deficit to the *QM*. *The electron dominance and deficit exist together.*

We denote elements, in accordance with the periodic table by $(n+1)$, (n) and $(n-1)$ in the sequence of the transformation of the matter in Fig.19.1 with proton mass $m_{p(n+1)}$, $m_{p(n)}$ and $m_{p(n-1)}$. The infinite cycle of the mass-energy transformation means that there should also be "indications" between the identified and measured elements. For addressing the cases we add and deduct a supposed Δm proton mass to and from the elements in the cycle.

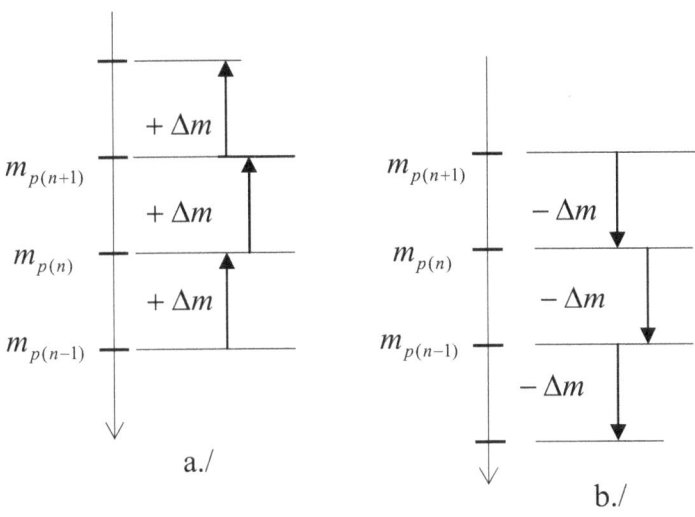

a./

Fig.19.1

Fig
19.1

It is easy to accept that if we add proton mass to elements (as is shown in Fig.19.1.a/), it will create electron dominance in each particular case, with blue shift surplus. With reference to 19F5, in spite of the added proton mass, the intensity of the neutron process stays unchanged, only its duration is longer. There is a blue shift surplus for use.

Should we deduct from each element a proton mass (as shown in Fig.19.1.b/), the result is the same unbalance, just the opposite: the process in each case goes with blue shift deficit. The red shift of the neutron collapse of elements needs more blue shift.

The quantum energy and the mass must be in balance. Therefore, the proton surplus and the proton deficit belong to all individual elements separately and act within the elements in parallel, as is shown in Fig.19.2.

The $(+\Delta m)$ and $(-\Delta m)$ chain covers the mass difference between each elements of the sequence of the mass-energy transformation.

The effects of the proton and neutron processes in each element are constant, since, in the case of surplus, less mass affects with more intensity; and, in shortfall, more mass with less intensity. The constancy of the intensity of the electron process at the same time guarantees the element.

The "added" proton mass results in electron dominance (with virtually more electrons) and more impact to the *Quantum Membrane*. The "deducted" proton mass results in less electron effect, in a virtual shortfall of electrons and in blue shift deficit.

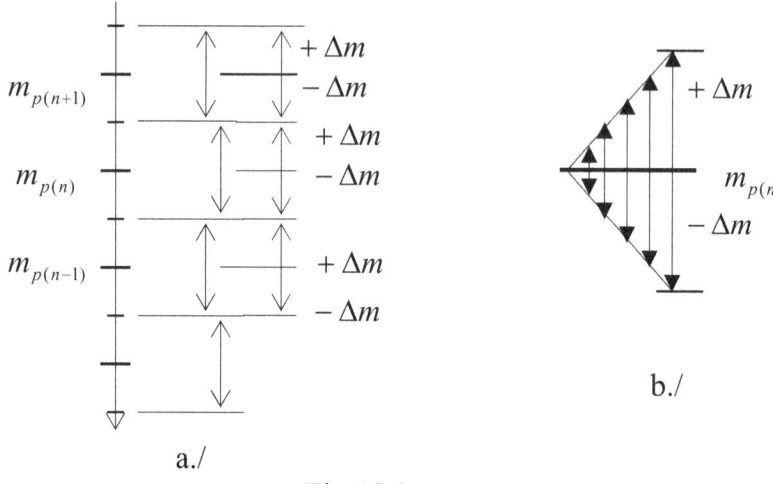

Fig
19.2

a./

Fig 19.2

Fig.19.2.b./ shows the constancy of element n with $m_{p(n)}$ proton mass. The summarised effect of the "above" (+) and "below" (-) proton masses guarantees the existence of the standard element.

Electrons are the end products of the proton transformation. The electron process itself ends with *quantum entropy*. More electron mass generates more (and less electron mass less) quantum entropy – the starting mass of the neutron collapse. The summarised effect of the blue shift surplus and deficit corresponds to the parameters of the element.

S.
19.1

19.1
Unbalanced electron process

We take a subject, which is built up from the atoms of the same element. The distinguishing feature of this subject is that the proton transformation, above *or* below the standard proton mass, has not been simultaneously balanced within the atoms of the element, as in normal structures.

This is still the same element, since the effect of the proton and neutron processes ("weights") correspond to the standard values and the electron intensity is constant. But at one *End* of the subject the atoms correspond to the structure with "added" proton mass and at the other *End* with "deducted" proton mass. The structure is smoothly changing from "added" to "deducted" between the two *Ends*. As a consequence of the proton mass, the electron weight is variable along the length of the subject: at the shortfall it is less, and at the dominant end it is more.

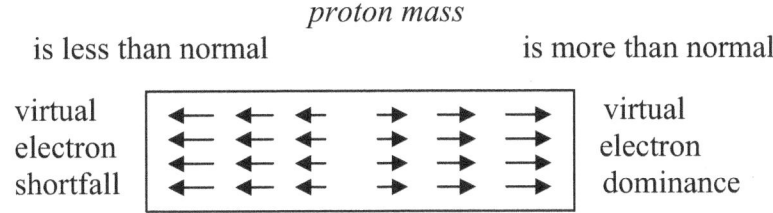

The arrows are proportional to the value of the "added" and "deducted" proton mass.

Fig.19.3

Fig
19.3

With reference to Section 19.0.1 and 19.0.2, the intensity of the neutron process in the subject is equal to normal. Its duration, however, is changing alongside the subject: from shorter to longer as the electron shortfall decreases and the electron dominance grows.

The two *Ends* differ in their blue shifts. The blue shift at the electron dominant *End* is more than required; at the *End* of the electron shortfall is not enough for the normal neutron process of the element.

Fig.19.3 represents the case: the changing energy balance structure within an element *in space* instead of in time. The energy balance is provided through the *Quantum Membrane*. We call the subject with unbalanced energy structure, a *magnet*.

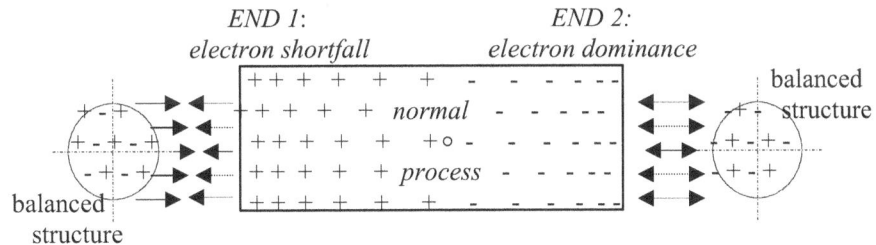

Fig.19.4

Fig.
19.4

19.2
The electron dominant *End* of the *magnet* (marked as 2)

Should we approach our magnet, with subjects *X* and *Y* with (normal) balanced energy structure (*changing in time*) from both ends, as shown in Fig.19.4, there will be different reactions between the photons and the electrons at the two *Ends*.

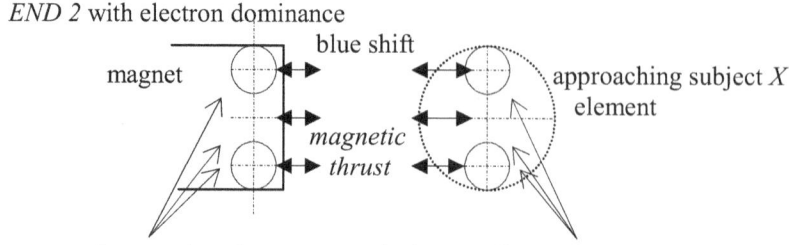

Fig
19.5
Fig.19.5

There is a *blue shift* conflict between the electrons of the electron dominant *END 2* of the magnet and the electrons of the subject on the approach to this *End*.

The electrons in both, the magnet and the subject are impacting the *Quantum Membrane* by *blue shift*. There is a difference between the two impacts. The blue shift of the electrons of the normal subject corresponds to normal energy balance conditions. The blue shift of the electrons in the magnet at this *End* is more than the *red shift* of the neutron collapse of the magnet at this *End* needs.

The blue shifts work against each other and prevent the subject of balanced energy status to approach the magnet at this *End*. The closer is the distance the greater is the thrusting effect between the electrons. The conflict either moves the electrons away from the conflict, or the magnet and the subject on approach will be in antagonistic magnetic thrust.

19.3.
The *End* of the *magnet* with electron shortage (marked as 1)

Because of the blue shift deficit, the neutrons of the element of the magnet at this *End* are without sufficient energy provision.

Subject *Y* on the approach is a blue shift source and there will be no blue shift conflict at this *End*. With reference to Section 18, the neutron collapse needs external energy to balance the growing mass of the incorporating photons. The magnetic attraction is a natural need.

The red shift formula of the magnet, with reference to 18A4 and 18A5 is:

$$f_v^{red} = f_i^{blue} \sqrt{1 - \frac{(c-v)^2}{c^2}}$$

The more is $u = c - v$, the speed difference of the collapse, the more is the effect of the blue shift of the approaching subject and the more is the *natural need* of the neutron collapse in energy.

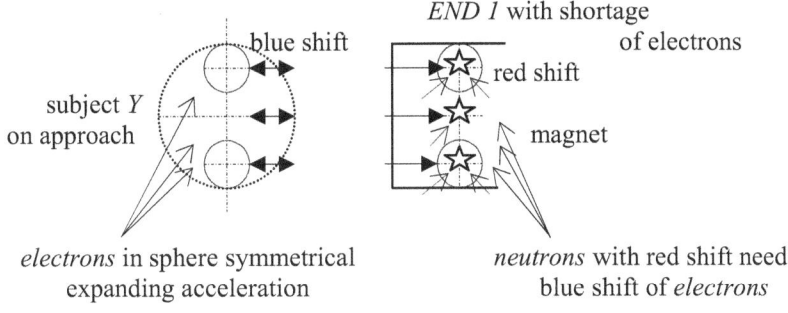

electrons in sphere symmetrical   neutrons with red shift need
expanding acceleration   blue shift of electrons

Fig.19.6

Fig
19.6

With reference to Fig.19.6, the missing blue shift to the red shift of the neutron collapse is provided by subject *Y* on the approach. Subject *Y* is attracted by *END 1* of the magnet.

As conclusion:
The energy and mass balance (of the element) of the magnet is changing in space rather than in time. This unbalanced status has impact on the surrounding environment through *magnetic field*. The magnetic field is the magnetic thrust and attraction existing in common.

The magnetic field acts and exists independently whether there is any subject in thrust or attraction. The magnetic field acts through the *Quantum Membrane* and provides the balance for the internal structure.

Its appearance depends on the energy structure and characteristics of the magnet itself and of the other subjects or elements of the magnetic impact. As events and experiments on ferromagnetism, diamagnetism or paramagnetism demonstrate.

The basis of the effect is the quantum energy and mass balance of the matter, existing in time.

The magnetic features relate mainly to elements and materials with blue shift deficit, but with event concentration value, close to $Z = 1$.

The ones with blue shift surplus are experiencing the conflict and mainly are to be found in gaseous status. Added or deducted proton values will not change significantly their (quasi-magnetic) energy-mass behaviour.

Elements with significant blue shift deficit can withstand the consequence of the change in the proton mass. A limited proton mass change will not change their general feature. They will be not sensitive even to higher intensity external blue shift impact. They remain stable and non-magnetised. (Elements with high atomic weight – and blue shift deficit – would need enormous impact to increase their blue shift surplus.)

Elements with blue shift deficit, close to $Z = 1$ are sensitive to magnetic effect. The added proton mass in these cases results in blue shift surplus, while the element or material stays with the same structure. The electron blue shift surplus of the *Carbon* (with $Z = 0.9868$) compensates the electron blue shift deficit of the *Iron* (with $Z = 1.1319$). The *Carbon Steel* is ideal for magnetising.

20

Quantum energy and mass balance of the elements of the periodic table

$\mathbf{T}$he energy mass balance analysis in Section 17 was applied to the full table of the periodic system of elements. The results are shown in Table 20.1 below.

Ref.
S.17

Element	Periodic Number P	M Atomic Weight measured	P Proton Mass	N Neutron Mass	Z Event Concentration of the Element
Hydrogen	1	1.0079	1.0072	0.00008	**0.000081**
Helium	2	4.0026	2.0145	1.9869	**0.9863**
Lithium	3	6.9400	3.0218	3.9165	**1.2961**
Beryllium	4	9.0120	4.0290	4.9807	**1.2362**
Boron	5	10.8100	5.0363	5.7709	**1.1458**
Carbon	6	12.0110	6.0436	5.9640	**0.9868**
Nitrogen	7	14.0067	7.0508	6.9519	**0.9860**
Oxygen	8	15.9990	8.0581	7.9364	**0.9849**
Fluorine	9	18.9984	9.0654	9.9280	**1.0951**
Neon	10	20.1700	10.0727	10.0918	**1.0019**
Natrium	11	22.9890	11.0799	11.9030	**1.0743**
Magnesium	12	24.3050	12.0872	12.2111	**1.0102**
Aluminium	13	26.9815	13.0945	13.8798	**1.0560**
Silicon	14	28.0855	14.1017	13.9760	**0.9911**
Phosphorus	15	30.9737	15.1090	15.8564	**1.0495**
Sulfur	16	32.0600	16.1163	15.9349	**0.9887**
Chlorine	17	35.4530	17.1235	18.3200	**1.0699**
Argon	18	39.9480	18.1308	21.8072	**1.2028**
Kalium	19	39.0980	19.1381	19.9494	**1.0424**
Calcium	20	40.0800	20.1454	19.9236	**0.9890**
Scandium	21	44.9559	21.1526	23.7917	**1.1248**
Titanium	22	47.9000	22.1599	25.7280	**1.1610**
Vanadium	23	50.9415	23.1672	27.7616	**1.1983**

Chromium	24	51.9960	24.1744	27.8083	**1.1503**
Manganese	25	54.9380	25.1817	29.7425	**1.1811**
Iron	26	55.8470	26.1890	29.6437	**1.1319**
Cobalt	27	58.9332	27.1962	31.7221	**1.1664**
Nickel	28	58.7100	28.2035	30.4911	**1.0811**
Cuprum	29	63.5400	29.2108	34.3132	**1.1746**
Zinc	30	65.3800	30.2181	35.1454	**1.1631**
Gallium	31	69.7350	31.2253	38.4926	**1.2327**
Germanium	32	72.5900	32.2326	40.3398	**1.2515**
Arsenic	33	74.9216	33.2399	41.6636	**1.2534**
Selenium	34	78.9600	34.2471	44.6941	**1.3050**
Bromine	35	79.9040	35.2544	44.6303	**1.2659**
Krypton	36	83.8000	36.2617	47.5185	**1.3104**
Rubidium	37	85.4658	37.2689	48.1765	**1.2927**
Strontium	38	87.6200	38.2762	49.3229	**1.2886**
Yttrium	39	88.9059	39.2835	49.6010	**1.2626**
Zirconium	40	91.2200	40.2908	50.9072	**1.2634**
Niobium	41	92.9064	41.2980	51.5858	**1.2491**
Molybdenum	42	95.9400	42.3053	53.6116	**1.2672**
Technetium	43	98.9620	43.3126	55.6258	**1.2843**
Ruthenium	44	101.0700	44.3198	56.7260	**1.2799**
Rhodium	45	102.9055	45.3271	57.5536	**1.2697**
Palladium	46	106.4000	46.3344	60.0403	**1.2958**
Silver	47	107.8680	47.3416	60.5005	**1.2779**
Cadmium	48	112.4100	48.3489	64.0347	**1.3244**
Indium	49	114.8200	49.3562	65.4369	**1.3258**
Tin	50	118.6900	50.3635	68.2991	**1.3561**
Antimony	51	121.7500	51.3707	70.3512	**1.3695**
Tellurium	52	127.6000	52.3780	75.1934	**1.4356**
Iodine	53	126.9045	53.3853	73.4901	**1.3766**
Xenon	54	131.3000	54.3925	76.8778	**1.4134**
Caesium	55	132.9054	55.3998	77.4754	**1.3985**
Barium	56	137.3300	56.4071	80.8921	**1.4341**
Lanthanum	57	138.9055	57.4143	81.4598	**1.4188**
Cerium	58	140.1200	58.4216	81.6665	**1.3979**
Praseodymium	59	140.9077	59.4289	81.4464	**1.3705**
Neodymium	60	144.2400	60.4362	83.7709	**1.3810**
Promethium	61	*145*	61.4434	83.5231	**1.3593**
Samarium	62	150.4000	62.4507	87.9152	**1.4077**
Europium	63	151.9600	63.4580	88.4674	**1.3942**
Gadolinium	64	157.2500	64.4652	92.7496	**1.4387**
Terbium	65	158.9254	65.4725	93.4172	**1.4268**

Dysprosium	66	162.5000	66.4798	95.9840	**1.4438**
Holmium	67	164.9304	67.4870	97.4065	**1.4433**
Erbium	68	167.2600	68.4943	98.7283	**1.4414**
Thulium	69	168.9342	69.5016	99.3947	**1.4301**
Ytterbium	70	173.0400	70.5089	102.4927	**1.4536**
Lutetium	71	174.9670	71.5161	103.4119	**1.4460**
Hafnium	72	178.4900	72.5234	105.9271	**1.4606**
Tantalum	73	180.9478	73.5307	107.3770	**1.4603**
Tungsten	74	183.8500	74.5379	109.2714	**1.4660**
Rhenium	75	186.2070	75.5452	110.6206	**1.4643**
Osmium	76	190.2000	76.5525	113.6058	**1.4840**
Iridium	77	192.2200	77.5597	114.6180	**1.4778**
Platinum	78	195.0900	78.5670	116.4802	**1.4826**
Gold	79	196.9665	79.5743	117.3488	**1.4747**
Mercury	80	200.5900	80.5816	119.9645	**1.4887**
Thallium	81	204.3700	81.5888	122.7367	**1.5043**
Lead	82	207.8000	82.5961	125.1589	**1.5153**
Bismuth	83	208.9804	83.6034	125.3315	**1.4991**
Polonium	84	*209*	84.6106	124.3432	**1.4695**
Astatine	85	*210*	85.6179	124.3354	**1.4522**
Radon	86	*222*	86.6252	135.3276	**1.5622**
Francium	87	*223*	87.6324	135.3198	**1.5417**
Radium	88	226.0254	88.6397	137.3374	**1.5494**
Actinium	89	*227*	89.6470	137.3042	**1.5316**
Thorium	90	232.0381	90.6543	141.3344	**1.5590**
Protactinium	91	231.0359	91.6615	139.3244	**1.5200**
Uranium	92	238.0290	92.6688	145.3097	**1.5680**
Neptunium	93	237.0482	93.6761	143.3211	**1.5300**
Plutonium	94	*244*	94.6833	149.2651	**1.5765**
Americium	95	*243*	95.6906	147.2572	**1.5389**
Curium	96	*247*	96.6979	150.2494	**1.5538**
Berkelium	97	*247*	97.7051	149.2416	**1.5275**
Californium	98	*251*	98.7124	152.2338	**1.5422**
Einsteinium	99	*254*	99.7197	154.2260	**1.5466**
Fermium	100	*257*	100.7270	156.2182	**1.5509**
Mendelevium	101	*258*	101.7342	156.2103	**1.5355**
Nobelium	102	*259*	102.7415	156.2025	**1.5293**
Lawrencium	103	*256*	103.7488	152.1947	**1.4670**
Rutherforium	104	*261*	104.7560	156.1869	**1.4909**
Dubnium	105	*262*	105.7633	156.1791	**1.4767**
Seaborgium	106	*263*	106.7706	156.1712	**1.4627**
Bohrium	107	*262*	107.7778	154.1634	**1.4304**

Hassium	108	265	108.7851	156.1556	**1.4354**
Meitnerium	109	266	109.7924	156.1478	**1.4222**
Darmstadtium	110	271	110.7997	160.1400	**1.4453**
Roentgenium	111	272	111.8069	160.1322	**1.4322**
Forecast	112	285	112.8142	172.1243	**1.5373**
Forecast	113	284	113.8215	170.1165	**1.4945**
Forecast	114	289	114.8287	174.1087	**1.5162**
Forecast	115	288	115.8360	172.1009	**1.4857**
Forecast	116	292	116.8433	175.0931	**1.4985**

Table
20.1   Table 20.1

Z in Table 20.1 is the *event concentration* of elements. With reference to 17M4, it is the quotient of the effects of the *neutron* and the *proton* change, equal to the quotient of the intensities of the *neutron* and *proton* processes. Z is one of the energy characteristics of the element. It shows the relative energy demand of the neutron process.

Ref
17M4

$$Z_{element} = \frac{N}{P} = \left| \frac{\varepsilon_n}{\varepsilon_p} \right|$$

The less the value of the event concentration, the more is the internal energy reserve of the element. At $Z=1$ the intensities of the neutron and the proton processes are equal.

M – is the atomic weight of the element, with reference to the periodic table, equal to the weight of neutrons, protons and electrons;

P – is the mass of the protons of the element, calculated from the periodic number and a mass of a single proton;

e – is the mass of the electrons of the atom of the element, calculated from the periodic number and the mass of a single electron;

N – is the mass of the neutrons within the atom of the element calculated as

20A1 $N = M - (P + e)$;

Mass of a single proton is: *1.00727 u* or $1.67262171 \cdot 10^{-27} kg$

Mass of a single electron is: *0.00055 u* or $9.10938215 \cdot 10^{-31} kg$

Ref   where u is the *unified atomic mass*

S.17.2   With reference to Section 17.2, the relation of N and P in 17M4 is equivalent to the weight relations: the effects of the mass change of the expanding acceleration of the protons and the collapse of the neutrons for infinite period of time. They are measured within the system of reference of the observation (*Earth*).

The effects of the neutron and proton processes give work value dimension.

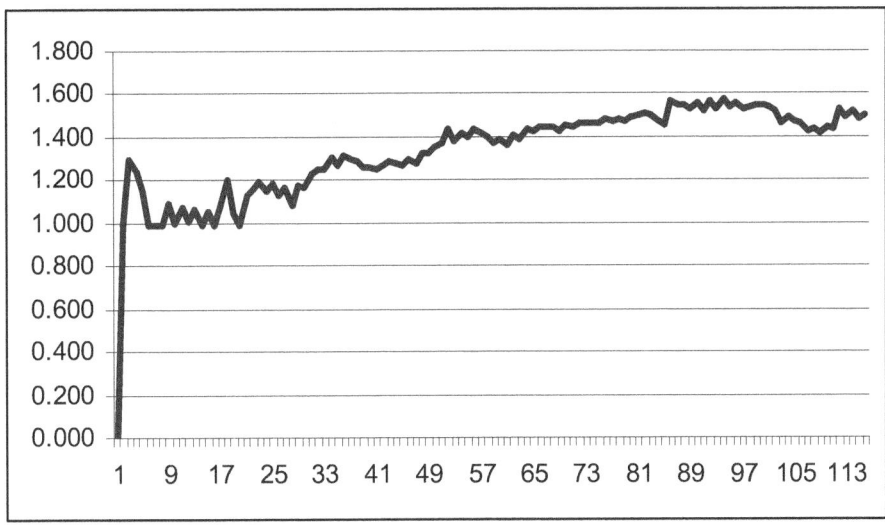

Diag.20.1

Diagram 20.1 shows the values of the event concentration for all elements of the periodic table. The horizontal axis shows the periodic number. The diagram shows divergences in some places and at the end of the curve. The reason for this could be the preciseness of the measurement and the forecast of the atomic weight at the end of the periodic table.

With reference to 18D4, the intensity of the electron process is:

$$\varepsilon_e = \frac{1}{Z}\sqrt{1 - \frac{(c-i)^2}{c^2}} \; ; \qquad \varepsilon_e = \left|\frac{\varepsilon_p}{\varepsilon_n}\right|\sqrt{1 - \frac{(c-i)^2}{c^2}}$$

The less the value of Z, the event concentration of the element is, the higher is ε_e, the intensity of the electron blue shift. The higher the blue shift is, the more "aggressive" is the element in reaction with others.

18D4 characterises the relation of the *intensity of the energy generation* (by proton) and the *intensity of the energy use* (by neutron) of the element.

Z, the event concentration within Table 20.1 grants specific features to the elements. These will be listed in the following sections.

20.1
Hydrogen

Hydrogen ($_1^1 H$) is *unique*. The extremely low value of Z means the extremely low *intensity* of the blue shift need of the neutron collapse. The extremely low intensity of the neutron process makes it possible to use the available blue shift of the *Hydrogen* for the red shift needs of the neutron collapse of other elements.

(The duration of the neutron collapse of the *Hydrogen* is expected to be infinitely long.)

$Z_H = 0.000081$, the value of the event concentration in this case means that the *single* electron of the *Hydrogen* is equivalent to the blue shift effect of

20A3
$$\varepsilon_{eH} = (n) = \frac{1}{0.000081}\sqrt{1 - \frac{(c-i)^2}{c^2}} \approx \frac{1}{0.000081} = 12345.6 \text{ electrons.}$$

Ref
18C4
With reference to 18C4, the effect of the electron process of the *Hydrogen* results in extremely big value of measured electron weight:

20B1
$$W_e = \frac{dm_e c^2}{dt_{sH}\varepsilon_{sH}} = \frac{\dot{m}_{eH}c^2}{\varepsilon_{sH}}\sqrt{1 - \frac{i^2}{c^2}}\left(1 - \sqrt{1 - \frac{(c-i)^2}{c^2}}\right) = n \cdot q = const$$

m_e in 20B1 is the unified electron mass. It is taken as equal for each element. The absolute work of the change of the unified electron mass keeps balance with equal number of photons for each element.

20B1 is valid for the *Hydrogen* only if the intensity of the mass change in the numerator is equivalent to the intensity of the mass change of the electron in the denominator:

Ref
20A2
$$\varepsilon_{sH} = \frac{1}{Z_H}\sqrt{1 - \frac{(c-i)^2}{c^2}}$$

Ref
18F5
With reference to 18F5, the weight is the effect of the mass change. For keeping the balance with the standard number of photons in 20B1, the intensity of the mass change (*the weight*) of the electron of the *Hydrogen* is proportional to the extremely high value of ε_{sH}.

20B2
$$w_e = \dot{m}_{eH}c^2\sqrt{1 - \frac{i^2}{c^2}}\left(1 - \sqrt{1 - \frac{(c-i)^2}{c^2}}\right)$$

The correct weight formula, from the point of view of the system of reference of the observation (in motion with lim$i=c$ relative to the system of reference of the electron), because of the intensity difference between the systems of reference, is:

20B3
$$w_{e-obs} = \dot{m}_{eH}c^2\frac{\sqrt{1 - (i^2/c^2)}}{\sqrt{1 - (i^2/c^2)}}\left(1 - \sqrt{1 - \frac{(c-i)^2}{c^2}}\right)$$

S.
20.2

20.2
Seven other specific elements

Oxygen ($_{15.9}^{8}O$), *Nitrogen* ($_{14.0}^{7}N$), *Helium* ($_{4}^{2}He$), *Carbon* ($_{12}^{6}C$), *Sulfur* ($_{32}^{16}S$), *Calcium* ($_{40}^{20}Ca$) and *Silicon* ($_{28}^{14}Si$).

The $Z < 1$ of these elements means: the intensity of the energy generation of the protons is more than the intensity of the energy utilization of the neutrons. (The intensity of the blue shift of the electrons is more than the intensity of the red shift of the neutrons.) Together with the *Hydrogen*, these are the *most* energy efficient elements: blue shift *"energy" providers*.

Elements with $Z < 1$ are keen to step into reaction with others and use their blue shift intensity "surplus" for intensifying the neutron process of other elements with $Z > 1$. The common mass energy balance of the molecules of the established compounds is based on the characteristics of the composing elements.

The external blue shift impact of elements with $Z < 1$ (most importantly *Oxygen, Nitrogen, Helium, Carbon*) may result in blue shift conflict with elements or compounds with balanced mass-energy structure.

As a general rule, the less Z, the event concentration, is, the more energy efficient the element is. The most energy efficient 16 elements are listed in Table 20.2. The values are based on recorded measured mass data, sensitivity of $10^{-27}\,kg$ for the *protons,* and $10^{-31}\,kg$ for the *electrons*. Therefore, Table 20.2 contains also those elements, the event concentration of which are close to $Z \geq 1$.

Element	PN	*M*	*Z*
Hydrogen	1	1.00790	**0.000081**
Oxygen	8	15.99900	**0.984897**
Nitrogen	7	14.00670	**0.985971**
Helium	2	4.00260	**0.986312**
Carbon	6	12.01100	**0.986841**
Sulphur	16	32.06000	**0.988744**
Calcium	20	40.08000	**0.988992**
Silicon	14	28.08550	**0.991084**
Neon	10	20.17000	1.001898
Magnesium	12	24.30500	1.059976
Kalium	19	39.09800	1.042393
Phosphorus	15	30.97370	1.049466
Aluminium	13	26.98150	1.059976
Chlorine	17	35.45300	1.069875
Natrium	11	22.98900	1.074281
Nickel	28	58.71000	1.081108
Fluorine	9	18.99840	1.095154

Table 20.2

Table
20.2

Table 20.2 contains the most active elements, which enter energy balance relations (chemical reaction) with others easily and widely.

20.3
Solid, liquid, gaseous and plasma phases

External blue shift impact intensifies the neutron and proton processes. It intensifies the internal blue shift conflict. If the intensity of the electron process remains unchanged, the element and the compound remain the same.

The external blue shift impact increases the magnetic thrust and the movement of atoms and molecules. The more is the impacting blue shift (the external energy intake), the more is the motion (the temperature) of the molecules. The measured *temperature* is the characterisation of the internal motion of atoms of elements and molecules of compounds.

The normal – not impacted – status is the *solid* phase. *Liquid* and *gaseous* phases have internal blue shift conflict.

At the temperature of absolute *zero* all elements are in solid phase. In this imaginary status, there is no energy communication between the *Quantum Membrane* and the elements. Elements in this status are only and fully impacted by their own electron blue shift. In *solid* phase the proton-electron-neutron process of elements or molecules of compounds has not been impacted by external blue shift (there is no energy intake by the elements).

In the case of growing blue shift impact from the *Quantum Membrane,* at the boiling point of elements or compounds, the electron blue shift will be in conflict (in magnetic thrust) with the blue shift of other electrons, in and out of the boundaries of the element. This is the *gaseous* phase. The gaseous phase means intensive motion of atoms of elements and molecules of compounds.

Helium, Hydrogen, Oxygen and *Nitrogen* have liquid phase and boiling point at extremely low temperatures.

Elements with $Z < 1$, with blue shift surplus, are in permanent blue shift conflict with the *Quantum Membrane*. While others (even in blue shift rich environment) are in solid or liquid phases, these are gaseous.

A certain level of *Quantum Membrane* impact is always present. Therefore the normal state of elements and compounds varies. In conventional meaning the blue shift intake from the *Quantum Membrane* means heating and the release means cooling.

The growing impact from the *Quantum Membrane* intensifies the blue shift conflict. The *plasma* is the phase with extremely high blue shift conflict. The speed of atoms in the conflict is equal to $\lim i = c$.

The *plasma* phase is unavoidable at the end of the mass transformation: With reference to Table 20.1, both, the *Hydrogen* and the *Helium*, the last two elements of the chain have blue shift surplus. The *Hydrogen* at the end of the mass transformation, with its infinitely long neutron process, is accumulating. As consequence, its blue shift impact is growing. The increasing blue shift conflict intensifies the collision between *Hydrogen-Hydrogen*, *Helium-Helium* and *Hydrogen-Helium* atoms. The speed of the atoms within the conflict is increasing. It grows up to $\lim i = c$.

The infinite blue shift impact destroys the *Helium* element. The collision of the atoms of the conflict at speed $\lim i = c$ also destroys the *Hydrogen* atoms. The distraction of the atomic structures results in additional blue shift impact to the *Quantum Membrane*. The *Quantum Membrane* with the additional blue shift impact intensifies the collisions. The reason and the consequence are the same.

The distraction of atomic structures within the *plasma* generates infinitely high energy intensity values relative to stationary systems of reference.

20.3.1. Blue shift sensitivity and the hydrocarbon

Elements can be characterised by their "sensitivity", (by their intensity of reaction) to the blue shift impact from the *Quantum Membrane*.

Elements with $Z < 1$ are blue shift sensitive. They provide blue shift to the *Quantum Membrane* and are in liquid and gaseous phases without impact. *Hydrogen*, *Oxygen*, *Nitrogen* and *Helium* are blue shift impact sensitive elements.

Elements with $Z \geq 1$ are not blue shift sensitive. These elements do not provide blue shift to others. The neutron collapse is more intensive than the proton transformation. The internal blue shift is taken by the neutrons in full. Therefore, these elements reach their liquid and gaseous phases after a long external blue shift impact (energy intake).

Carbon is an element with $Z \cong 1$. The intensities of the proton and neutron processes are quasi equal. *Carbon* does not have blue shift impact to the *Quantum Membrane*. This is the explanation for its high temperature melting point and even higher boiling point temperature. *Carbon* however has different features in compound with *Hydrogen*.

Hydrocarbons are compounds with certain compositions of the *Carbon* and the *Hydrogen*. The *Hydrogen* has an infinitely long neutron collapse and exceptionally high blue shift surplus.

The blue shift of the electron process of the *Hydrogen* intensifies the neutron process of the *Carbon*. One single *Hydrogen* atom could not result in the formulation of *Hydrocarbons*. But there must be a certain number of *Hydrogen* atoms within the molecule to result in *Hydrocarbons*.

The neutron process of the *Carbon* will be increased. It will result in internal blue shift conflict, which will result in increased motion of the *Hydrocarbon* molecules. The *Hydrocarbons* are liquids and sensitive to external blue shift. With additional external blue shift impact *Hydrocarbons* can easily reach the gaseous stage.

S.
20.3.2   *20.3.2. Water*

The key here is the effect of the blue shift surplus of the *Oxygen*.

All elements except *Hydrogen* are destroyed in reaction with *Oxygen*. The blue shift surplus of the *Oxygen* speeds up the neutron process of the *Hydrogen*. It however remains infinitely slow even after the impact.

The result is *Water*, full of blue shift surplus – energy. The molecules of the *Water* are in conflicting internal blue shift with each other. The blue shift surplus of the molecules of the *Water* however is less than the blue shift surplus of the two elements separately.

S.
20.4
20.4
Beta decay, gamma and alpha radiation of elements

Ref
S.18.2   The proton-electron-neutron processes within the atoms of elements are in
Ref   balance. Should this balance be destroyed, the consequence – in our
Table   conventional understanding – is radiation, a certain process of the element
18.1   for survival.

The key factors of the energy balance are: the *event concentration* and the *intensity relation* of the proton-electron-neutron process. These are constant and standard for each element:

20C1
$$\varepsilon_e = \frac{\varepsilon_p}{\varepsilon_n}\sqrt{1 - \frac{(c-i)^2}{c^2}} = const \quad \text{and} \quad Z = \frac{N}{P} = \left|\frac{\varepsilon_n}{\varepsilon_p}\right| = const$$

Indexes *e*, *p*, and *n* to intensity ε in 20C1 denote the electron, proton and neutron processes.

S.
20.4.1   *20.4.1. Beta decay and gamma radiation*

If the proton process is unbalanced, it will result either in extra or missing electrons within the element.

In the case of extra electron unbalance, the blue shift of the electron surplus must be released, resulting in so-called β *decay*.

If the proton mass unbalance results in missing electron, with reference to 16D1-16D6 and Section 18.2, the neutron process becomes unstable and the collapse can partially stop and turn into expanding acceleration.

Ref 16D1- 16D6 S.18.2

With reference to 19E1, the electron blue shift provides the energy of the neutron collapse:

$$dmc^2\sqrt{1-\frac{i^2}{c^2}}\left(1-\sqrt{1-\frac{(c-i)^2}{c^2}}\right)=\frac{dmc^2\sqrt{1-\frac{i^2}{c^2}}\sqrt{1-\frac{(c-i)^2}{c^2}}}{\sqrt{1-\frac{(c-v)^2}{c^2}}}-dmc^2\sqrt{1-\frac{i^2}{c^2}}\sqrt{1-\frac{(c-i)^2}{c^2}}$$

Ref 19F1

The frequency of the collision of the photons with the accelerating (instead of collapsing) neutron depends on the actual speed of the collapse (between $v = \lim 0$ and $v = i$). The lower is the speed the more photons are collision for the unit period of time. The frequency will be *blue shifted* rather than *red shifted* by the *accelerating* (instead of collapsing) neutron. It will result in γ decay – high frequency impact to the *Quantum Membrane*.

20.4.2. Alpha radiation

S. 20.4.2

Should the proton-electron-neutron processes result in mass unbalance, which needs mass release, the element will be safe from destruction with an α - decay.

The "weight" of the element is

$$w = \left(\dot{m}_p c^2 + \dot{m}_p c^2 \sqrt{1-\frac{(c-i)^2}{c^2}} \right) \cdot \left(1-\sqrt{1-\frac{i^2}{c^2}} \right)$$

20C2

The neutron and the proton mass relation is $\quad m_n = m_p \sqrt{1-\frac{(c-i)^2}{c^2}}$

20C3

The *alpha decay* is the release of the smallest possible mass from the structure of the element.

The smallest existing element is the *Hydrogen*, but the intensity of the neutron process of the *Hydrogen* is infinitely low and the process itself is infinitely long. Therefore, $_1^2H$, an element as such, in fact does not exist. The mass correction cannot be made by releasing a mass corresponding to the weight of the *Hydrogen*. The smallest energy *sorting-out-portion* of the imbalance is the nucleus of the atom of the $_2^4He$ element.

The proton is the result of the neutron process. Without the fully collapsed neutron in the cycle there is no proton in the next cycle. Each proton cycle means a different type of matter – with reference to 17E5, a *different element* with less measured mass value of the proton by:

Ref 17E5

$$\Delta m = m\left(1-\sqrt{1-\frac{(c-i)^2}{c^2}} \right)$$

20C4

Ref
17E7

With reference to 17E7, after an infinite number of cycles, the process arrives at the smallest possible element, $_1^1H$, *Hydrogen*, without a formulated neutron, which as per the law of the natural entropy, cannot terminate itself by turning into a proton. The disappearance of the neutron of the *Hydrogen* would mean the "end" of the process, which obviously cannot be the case. Our mass measurement, however, gives a figure very close to zero. In fact, the event concentration of *Hydrogen* $_1^1H$ and the mass of

20C5

the neutron of the *Hydrogen* will be infinitely small, approaching zero: $\lim Z_{_1^1H} = 0$

Atomic mass: $M_{_1^1H} = 1.00790\ u;$ Mass of the proton: $P_{_1^1H} = 1.00727\ u$

Mass of the electron: $e_{_1^1H} = 0.00055\ u$

Mass balance of the neutron: $N_{_1^1H} = M_{_1^1H} - (P_{_1^1H} + e_{_1^1H}) = 0.000081\ u$

S.
20.4.3

20.4.3. *Change into new element*

Once the proton and neutron energy-mass (and time) balance of an element is disrupted, the modified new energy-mass balance will create a new element in accordance with the energy-mass (and time) structure.

The principle of the balance is based on
- the event concentration of the neutron-proton processes, and
- the energy-mass and time balance of the proton transformation, the electron blue shift and the neutron red shifts.

21

Quantum engines

21.1
The electricity generator and electric engine

The electrons within the electromagnet of the *stator* and within the wires of the *rotor* of the electricity generator are in blue shift conflict: the stator and the rotor of the generator are in magnetic thrust and the rotor cannot rotate freely within the stator. External work must be spent for rotating it, as much as the capacity (the unbalance) of the electromagnet. The wires can be rotated together with the wheel, but the electrons cannot move. They will flow away within the wires of the solenoid coil of the rotor to avoid the blue shift conflict. The generator produces *electricity*. The electricity has its certain frequency, depending on the rotation of the wheel of the generator.

In the electric engine the process is similar, just the inverse to the one in the generator: the moving electrons, "the electricity" within the wires of the rotor, move the electrons into blue shift conflict with the electrons of the electron-dominant end of the permanent magnet. The only way the engine can resolve the conflict is by pushing the wheel into rotation.

The electromagnet is the quantum engine which does not have moving components. The wire and the iron core are fixed together. The blue shift of electrons passing through within the coil of the wires moves the electrons in one direction within the iron core of the magnet. The accumulating increased number of electrons at one end will create electron dominance. The other end, at the same time, will be short of electrons.

The "quantum engine electromagnet" is at rest and the electrons are moving inside to avoid the blue shift conflict. The electromagnet has electron dominance and electron shortage until external energy moves the electrons within the wire.

The blue shift conflict is the principle of the operation. In order to sort the conflict out the generator generates electron-flow, the engine rotates the wheel and the electron flow creates the electromagnet.

S.
21.2

21.2

The *Hydrocarbon* (petrol) engine and *Fire*

The electron blue shift of the *Hydrogen* is the most intensive among the elements. It intensifies the red shift of the neutron collapse in the molecules of *Hydrocarbon* compounds. The mix with *Air* (O_2 and N_2) adds additional blue shift to the liquid *Hydrocarbon* compound. With reference to Section 20.2, O_2 and N_2 both are aggressive blue shift providers and their additional blue shift threatens the *Hydrocarbon* molecule with destruction.

Ref
S.20.2

With reference to 17L3, the neutron and proton processes for a balanced element are:

21A1
$$\frac{dmc^2}{dt_p \varepsilon_p}\left(1-\sqrt{1-\frac{i^2}{c^2}}\right) = \frac{dmc^2}{dt_n \varepsilon_n}\sqrt{1-\frac{(c-i)^2}{c^2}}\left(\sqrt{1-\frac{i^2}{c^2}}-1\right)$$

21A2
$$\frac{\dot{m}_p c^2}{\varepsilon_p}\left(1-\sqrt{1-\frac{i^2}{c^2}}\right) = \frac{\dot{m}_n c^2}{\varepsilon_n}\sqrt{1-\frac{(c-i)^2}{c^2}}\left(\sqrt{1-\frac{i^2}{c^2}}-1\right)$$

With reference to Section 18.2, the electron process in balanced conditions is:

21A3
$$\frac{dmc^2}{dt_e \varepsilon_e}\sqrt{1-\frac{i^2}{c^2}}\left(1-\sqrt{1-\frac{(c-i)^2}{c^2}}\right) = \frac{dn}{dt_e \varepsilon_e}q = \frac{f_e}{\varepsilon_e}q = const$$

Since the frequency of the blue shift of the photons corresponds to the constant speed of the sphere symmetrical expanding acceleration, the motion with $i = \lim a\Delta t = c = const$, the electron processes in absolute terms for all elements are equal and constant. 21A3 shows that with the increase of the intensity of the electron process, the constancy could only be granted if the frequency is increased.

For the system of reference of a certain element with ε_e - electron process intensity and $\dot{m}_e$ - electron mass change, the frequency would be:

21A4
$$\dot{m}_e c^2\left(1-\sqrt{1-\frac{(c-i)^2}{c^2}}\right) = f_e \cdot q$$

From 21A1 it follows that if

21A5 $\varepsilon_e = 1$ the frequency is f_i, and it corresponds to $\lim i = c$

In the case of $\varepsilon_e > 1$ (or for the element itself: $Z < 1$),

21A6   the frequency would be $f_e > f_i$.

For the *Hydrogen,* with reference to Section 20, it would be infinitely high. For all elements with *blue shift surplus*, it would be more than f_i.

In the case of $\varepsilon_e < 1$ (or for the element itself: $Z > 1$), for all elements with

21A7   electron blue shift deficit it would be $f_e < f_i$.

The blue shift surplus of the *Oxygen* and *Nitrogen* increases the intensity of the neutron red shift of the *Hydrocarbon* molecules. Within the

Hydrocarbon molecules the neutron process of the *Carbon* has already been intensified by the *Hydrogen*. The intensified neutron process will result in intensified proton process, since:

$$\varepsilon_e^{Hydrocarbon} = \frac{\varepsilon_p^{Hydrocarbon}}{\varepsilon_n^{Hydrocarbon}}\sqrt{1 - \frac{(c-i)^2}{c^2}} \qquad 21B1$$

The permanent *Oxygen* and *Nitrogen* blue shift impact results in infinite intensity of the neutron process and neutron mass change within the *Hydrocarbon* molecules: $\lim \dot{m}_n = \infty$ and $\lim \varepsilon_n = \infty$

This obviously will intensify the proton process of the *Hydrocarbon* to infinite value: $\lim \dot{m}_p = \infty$ and $\lim \varepsilon_p = \infty$

There is no formula to describe the process with infinite intensity values:

$$Z = \left| \frac{\varepsilon_n}{\varepsilon_p} \right| = \frac{\infty}{\infty} \qquad 21B2$$

The *Hydrocarbon* molecules will be destroyed.

S.
21.2.1

21.2.1 Fire

The electron process of the destroyed *Hydrocarbon* molecules is an impact on the *Quantum Membrane*. This destruction of atoms of elements or molecules of compounds – from the point of view of the system of reference of the observation *(Earth)*, is *Fire: electron blue shift with infinite frequency*.

$$\frac{\dot{m}_e c^2 \sqrt{1 - \frac{i^2}{c^2}}}{\sqrt{1 - \frac{i^2}{c^2}}} \left(1 - \sqrt{1 - \frac{(c-i)^2}{c^2}} \right) = \frac{f_i}{\sqrt{1 - \frac{i^2}{c^2}}} q \qquad 21C1$$

The petrol engine burns the *Hydrocarbon* molecules. The *Fire* is electron blue shift impact on the *Quantum Membrane*: the release of *quantum energy* of infinite frequency.

The *Fire* in petrol engines, since it adds extra blue shift of infinite frequency to the already existing blue shift surplus of the *Hydrocarbon* molecules, can speed up the destruction process.

The *Oxygen* blue shift surplus is significant and the effect of the proton process results in 8 electrons. The blue shift surplus of 8 electrons - as external blue shift impact to elements or compounds with blue shift surplus - destroys the element and the compounds.

Water extinguishes *Fire*. *Water* molecules are in blue shift conflict and take more blue shift to getting into gaseous (steam) phase. The use of the blue shift surplus of the *Oxygen* stops the *Fire*. Liquid *Nitrogen* and *Helium* also extinguish *Fire*.

S.
21.3

21.3

Nuclear engine

The difference between $^{238}_{92}U$, the "normal" *Uranium* element and the $^{235}_{92}U$ isotope, is the intensity in their electron processes. $^{235}_{92}U$ is to be found in Nature in the concentration of 0.7% of the basic *Uranium* element.

Ref
18B5

With reference to 18B5 the intensity relation of the proton-electron-neutron processes characterize the element

21D1

$$\varepsilon_e = \frac{\varepsilon_p}{\varepsilon_n} \sqrt{1 - \frac{(c-i)^2}{c^2}}$$

$^{235}_{92}U$ with its less neutron weight and with its equal to $^{238}_{92}U$ proton weight obviously has electron blue shift surplus relative to $^{238}_{92}U$.

> For good measure we have to remember that the measured weight (of the particles) is the effect of the mass change.
> For the proton process it is: and for the neutron process it is:

21D2
21D3

$$w_p = \dot{m}_p c^2 \left(1 - \sqrt{1 - \frac{i^2}{c^2}} \right); \qquad w_n = \dot{m}_n c^2 \sqrt{1 - \frac{(c-i)^2}{c^2}} \left(\sqrt{1 - \frac{i^2}{c^2}} - 1 \right);$$

> The higher the intensity of the mass change, the higher is the effect (and the measured weight).

Uranium $^{238}_{92}U$ is stable and the intensity relations are balanced. Its event concentration value is $Z_{U^{232}_{92}} = 1.5680$.

Ref
S.20

With reference to Section 20, $Z > 1$ means in general that the internal quantum energy mass balance goes with "blue shift deficit". The intensity of the proton process (and the electron process – the drive of the neutron collapse) is behind the intensity of the neutron mass change. From this point of view, the neutron process of $^{235}_{92}U$ goes with less blue shift deficit. Comparing it to $^{238}_{92}U$, the $^{235}_{92}U$ isotope has "blue shift surplus".

The 0.7% concentration of the $^{235}_{92}U$ isotope "communicates" with a small part of the basic *Uranium* element. Its naturally small proportion cannot alone cause more serious consequences. The blue shift may result in a limited increase in temperature of the surrounding *Uranium* atom environment.

The blue shift "surplus" of higher concentration of the $^{235}_{92}U$ isotope within the $^{238}_{92}U$ element either

-   causes massive β-decay, which itself alone would increase the intensity of the neutron process of the $^{238}_{92}U$ element; or

- splits the isotope into fission products and blue shift. (Because the intensity of the electron process and the event concentration of the $^{235}_{92}U$ isotope do not correspond to those of the $^{238}_{92}U$ element.)

With reference to the findings in Section 20, the *Hydrogen, Oxygen* and *Carbon* elements are among the biggest blue shift providers in Nature. The fuel of nuclear reactors is the $^{238}_{92}U$ element in various enrichment of the $^{235}_{92}U$ isotope.

The $^{235}_{92}U$ fission and the additional blue shift, provided by the atoms of the *Water* (in reactors with *water* moderator,) or *Graphite* (in reactors with *graphite* moderator) result in the increase of the intensity of the neutron collapse of the atoms of $^{238}_{92}U$. The more intensive neutron process results in a more intensive proton process. The electron process of the increased neutron and proton processes remains without change:

$$\varepsilon^{238}_{92e} = \frac{\varepsilon^{238}_{92p}}{\varepsilon^{238}_{92n}}\sqrt{1 - \frac{(c-i)^2}{c^2}} = \frac{\varepsilon^{239}_{92p}}{\varepsilon^{239}_{92n}}\sqrt{1 - \frac{(c-i)^2}{c^2}} = \varepsilon^{239}_{92e} \qquad \text{21E1}$$

Intensities ε^{238}_{92} and ε^{239}_{92} with index e, p and n denote the intensities of the electron, proton and neutron processes in the *Uranium* atoms with atomic weight 238 and 239.

The process - *without cooling* - stops, leaving the basic *Uranium* element in higher energy level (higher temperature). The higher intensity of the neutron and proton processes result in higher measured weight of the element and in speeded up proton-electron-neutron cycle, causing intensive internal blue shift conflict. (In conventional terms it means higher temperature. It is one of the safety features of specific nuclear reactors. The temperature increase slows down the chain reaction.)

The increase of the intensity of the neutron process results in higher effect ("weight") of the neutrons. The result of the neutron-proton balance results in the increase of the intensity of the proton process as well. (The common mass milestone guarantees the higher intensity.) Thus, in this case, the proton weight also increases. The new periodic number grows from 92 to 93.

But this is a new element: *Neptunium*.

$$^{239}_{92}U \rightarrow {}^{239}_{93}Np \qquad \text{21E2}$$

With reference to Table 20.1, the even concentration of the *Neptunium* $Z_{Np} = 1.5300$, is less than the event concentration of the basic *Uranium*, $Z_U = 1.5680$. The result is surplus in electron process intensity.

The correct description of 21E2 is: $^{239}_{92}U \rightarrow {}^{239}_{93}Np + \beta$ 21E3

where β means the effect of the available electron process:

21E4
$$\varepsilon_{93e}^{239} = \frac{\varepsilon_{93p}^{239}}{\varepsilon_{93n}^{239}} \sqrt{1 - \frac{(c-i)^2}{c^2}} > \varepsilon_{92e}^{238}$$

The process continues and the chain ends at *Plutonium* $_{94}^{239}Pu$, again with massive event concentration value of $Z_{Pu} = 1.5765$, more than that of the *Uranium*.

The function of the *Water* in nuclear reactors is not just for providing blue shift (moderator function), but also for taking off "energy" from the transforming matter, for cooling it.

S.
21.3.1   *21.3.1. Nuclear fission in nuclear reactors*

The nuclear fuel is based on $_{92}^{238}U$, the natural *Uranium* element in necessary enrichment with $_{92}^{235}U$ the *Uranium isotope* (simply *Uranium Fuel* or just *Fuel)* and the moderator is H_2O, (light) *Water*.

The blue shift of the *Uranium* in absolute terms is:

21F1
$$W_U^{blue} = \frac{\dot{m}_U^e c^2}{\varepsilon_U^e} \left(1 - \sqrt{1 - \frac{(c-i)^2}{c^2}} \right)$$

The blue shift surplus provided by the *Water* in absolute terms is:

21F2
$$W_{Water}^{blue} = \frac{\Delta \dot{m}_W^e c^2}{\varepsilon_W^e} \left(1 - \sqrt{1 - \frac{(c-i)^2}{c^2}} \right)$$

The blue shift of the $_{92}^{235}U$ isotope is:

21F3
$$W_{U235}^{blue} = \frac{\dot{m}_{U235}^e c^2}{\varepsilon_{U235}^e} \left(1 - \sqrt{1 - \frac{(c-i)^2}{c^2}} \right)$$

$\dot{m}_U^e$, $\dot{m}_{U235}^e$ and $\Delta \dot{m}_W^e$ are the values of the electron mass change of the *Uranium* element and the $_{92}^{235}U$ isotope and the surplus of the mass change of the *Water*; ε_U^e, ε_{U235}^e and ε_W^e are the intensities of the electron processes of the *Uranium*, the $_{92}^{235}U$ isotope and the molecules of the *Water*.

> The intensity of the blue shift of the *Hydrogen* and the *Oxygen* are different, but we take the blue shift of the *Water* molecule as an integrated common one.

With reference to 18G2, the effect of the summarized blue shift of the *Uranium*, the $_{92}^{235}U$ *isotope* (the components of the *Fuel*) and the surplus of the moderator-cooling *Water* is:

$$\left(\frac{\dot{m}_U^e c^2}{\varepsilon_U^e} + \frac{\dot{m}_{U235}^e c^2}{\varepsilon_{U235}^e} + \frac{\Delta \dot{m}_W^e c^2}{\varepsilon_W^e} \right) \sqrt{1 - \frac{i^2}{c^2}} \left(1 - \sqrt{1 - \frac{(c-i)^2}{c^2}} \right) =$$

$$= \frac{\dot{m}_{UF}^n c^2}{\varepsilon_{UF}^n}\left(1 - \sqrt{1 - \frac{(c-i)^2}{c^2}}\right) \qquad\qquad 21F4$$

Indexes *U* and *W* denote the *Uranium* and the *Water* accordingly.

$\dot{m}_{UF}^n$ and ε_{UF}^n in 21F3 denote the modified mass change intensity and the modified intensity value of the neutron process of the *Uranium Fuel*. The parameters are the result of the effect of the blue shift of the *Water* and the $^{235}_{92}U$ isotope as the moderators of the process.

The red shift of the neutron process of the *Uranium Fuel* on the right hand side of 21F4 is in balance with the summa blue shift of the electrons of the atoms of the *Uranium*, the $^{235}_{92}U$ *isotope* and the blue shift surplus of the *Water* molecules. This summarized blue shift on the left hand side of 21F4 is more than the natural blue shift of the $^{238}_{92}U$ element and the $^{235}_{92}U$ *isotope*. This is the drive for the increase of the internal blue shift conflict (temperature increase) of the $^{238}_{92}U$ element and the increasing unbalance of the $^{235}_{92}U$ *isotope* of the *Uranium Fuel*.

We do not know what the common integrated or average electron process intensity of the *Water* is, but with reference to Section 20, for the *Hydrogen* and for the *Oxygen* they are accordingly:

$$\varepsilon_H^e = \frac{1}{Z_H^e}\sqrt{1 - \frac{(c-i)^2}{c^2}} \approx 12345.68 \quad \text{and} \quad \varepsilon_O^e = \frac{1}{0.9849}\sqrt{1 - \frac{(c-i)^2}{c^2}} \approx 1.015 \qquad \begin{array}{l} 21F5 \\ 21F6 \end{array}$$

for the *Uranium*: $\quad \varepsilon_U^e = \frac{1}{Z_U^e}\sqrt{1 - \frac{(c-i)^2}{c^2}} = \frac{1}{1.5680}\sqrt{1 - \frac{(c-i)^2}{c^2}} \approx 0.6377 \qquad 21F7$

for the $^{235}_{92}U$ *isotope:* $\quad \varepsilon_{U\,235}^e = \frac{P_{235}}{N_{235}}\sqrt{1 - \frac{(c-i)^2}{c^2}} \approx 0.6434 \qquad \begin{array}{l} 21F8 \\ \text{Ref} \end{array}$

With reference to 17L3 on the equality of the proton and neutron processes: $\quad$ 17L3

$$\frac{\dot{m}_{UF}^p c^2}{\varepsilon_{UF}^p}\left(1 - \sqrt{1 - \frac{i^2}{c^2}}\right) = \frac{\dot{m}_{UF}^n c^2}{\varepsilon_{UF}^n}\sqrt{1 - \frac{(c-i)^2}{c^2}}\left(\sqrt{1 - \frac{i^2}{c^2}} - 1\right) \qquad 21G1$$

$\dot{m}_{UF}^p$ and ε_{UF}^p are the modified mass change of the transformation and the modified intensity of the proton process of the *Uranium Fuel*.

In absolute terms, the mass effect of the *Fuel* at the end of the proton process is equal to the mass effect of the *Fuel* of the electrons process. This blue shift is more than the genuine blue shift of the basic *Uranium* element.

$$W_{UF}^e = \frac{\dot{m}_{UF}^e c^2}{\varepsilon_{UF}^e}\left(1 - \sqrt{1 - \frac{(c-i)^2}{c^2}}\right); \qquad \text{and} \qquad \frac{\dot{m}_{UF}^e}{\varepsilon_{UF}^e} = \frac{\dot{m}_{UF}^p}{\varepsilon_{UF}^p}\sqrt{1 - \frac{i^2}{c^2}} \qquad \begin{array}{l} 21G2 \\ 21G3 \end{array}$$

The equality in 21G3 is the consequence of the common mass milestone.

In absolute terms the mass energy balance is granted: the increased blue shift of the *Uranium Fuel* should be equal to the blue shift, provided by the *Water-moderator* (or in other cases by the *Graphite-moderator*):

21G4
$$\Delta W_{UF}^{blue} = \left(\frac{\dot{m}_{UF}^e c^2}{\varepsilon_{UF}^e} - \frac{\dot{m}_U^e}{\varepsilon_U^e} - \frac{\dot{m}_{U235}^e}{\varepsilon_{U235}^e} \right) \left(1 - \sqrt{1 - \frac{(c-i)^2}{c^2}} \right) =$$

$$= \frac{\Delta \dot{m}_W^e}{\varepsilon_W^e} \left(1 - \sqrt{1 - \frac{(c-i)^2}{c^2}} \right)$$

The blue shift in 21G2, provided by the modified electron process of the atoms of the *Uranium Fuel* is *more* than the blue shift of the natural *Uranium* element and $^{235}_{92}U$ isotope. The *Uranium* element as a consequence of the increased internal blue shift conflict is warming up, but will not be destroyed, since its electron intensity will be not changed. The naturally unbalanced $^{235}_{92}U$ isotope, however, cannot keep its atomic structure. The effects of the blue shift surplus of the *Water* and the increased enrichment of $^{235}_{92}U$ within the *Fuel* will split the $^{235}_{92}U$ isotope into elements (fission products), with summarized electron process intensity, equal to the increased unbalance value the $^{235}_{92}U$ isotope of the *Uranium Fuel*:

21G5
$$\frac{\dot{m}_{U235}^e c^2}{\varepsilon_{U235}^e} \left(1 - \sqrt{1 - \frac{(c-i)^2}{c^2}} \right) = \left(\frac{\dot{m}_A^e c^2}{\varepsilon_A^e} + \frac{\dot{m}_B^e c^2}{\varepsilon_B^e} + ... \right) \left(1 - \sqrt{1 - \frac{(c-i)^2}{c^2}} \right)$$

The summarized mass of the *fission products* is less than the mass of the $^{235}_{92}U$ isotope. The event concentration values (Z) of each of the fission products are also less than that of the $^{235}_{92}U$ isotope. Therefore, the intensity of each of the electron processes of the fission products is more than the intensity of the $^{235}_{92}U$ isotope (and also, obviously, of the *Uranium*).

It means more frequent collisions as a result of the internal blue shift conflict within these elements. In conventional terms it means higher temperature. The resulting blue shift increase (temperature) is taken away by the cooling *Water* (and moderator).

In one's conventional understanding of the fission, the mass difference is transformed into energy. In fact, the blue shift of the $^{235}_{92}U$ isotope and the blue shift surplus (of the *Hydrogen* and the *Oxygen*) of the *Water* (as moderator) are transformed into blue shifts of fission product and into the increased internal blue shift conflict of the *Uranium* element.

And all this blue shift increase and blue shift conflict is removed by the *Water* (as cooling medium). These energies are perfectly equal.

$$\left(\frac{\Delta \dot{m}_W^e c^2}{\varepsilon_W^e} + \frac{\dot{m}_{U235}^e c^2}{\varepsilon_{U235}^e} \right) \left(1 - \sqrt{1 - \frac{(c-i)^2}{c^2}} \right) =$$

$$= \frac{\Delta \dot{m}_U^e}{\varepsilon_U^e} + \left(\frac{\dot{m}_A^e c^2}{\varepsilon_A^e} + \frac{\dot{m}_B^e c^2}{\varepsilon_B^e} + ... \right) \left(1 - \sqrt{1 - \frac{(c-i)^2}{c^2}} \right) \qquad \text{21G6}$$

The more intensive the impacting electron intensity of the moderator is, the more intensive is the neutron collapse of the *Uranium Fuel*. The increased neutron process results in more intensive proton process. The increased proton process corresponds to more intensive electron acceleration in the $_{92}^{235}U$ *isotope* and more intensive internal blue shift conflict of the *Uranium* element. The increased blue shift splits the $_{92}^{235}U$ *isotope* into fission products with summarized mass less than the basic element. The intensive blue shift of the fission products and the *Uranium* element results in internal blue shift conflict (more frequent collision) temperature increase, the "energy" from which is taken away by the *Water* cooling moderator.

Nuclear reactors with *Uranium Fuel* with natural content of $_{92}^{235}U$ work on the same principle. The only difference is that the moderator is *Deuterium* D_2O with more intensive blue shift, which is necessary because of the lower volume proportion of the $_{92}^{235}U$.

At a certain enrichment – critical mass – the $_{92}^{235}U$ isotope itself (without *Water*) can provide sufficient "moderator function" and destroy itself.

The natural *Uranium*, as a consequence of the blue shift provided by the $_{92}^{235}U$ isotope, is in energy exchange with the environment. (The tiny blue shift intensity surplus of $_{92}^{235}U$ isotope is taken away by the environment.) This ensures the stability of the heat exchange below the critical mass.

Above the critical mass, the natural cooling cannot release the increased energy of the blue shift of the $_{92}^{235}U$ *isotope* and uncontrolled chain reaction (fission) occurs. The uncontrolled chain reaction means the boundaries of the earlier systems of reference of the mass energy balance will be broken. The unbalanced process happens within the system of reference, stationary relative to the system of reference of the elements, with infinitely slow time flow. The stationary status relative to the system of reference of the electrons in motion with $i = \lim a \Delta t = c$ results in an event with infinitely high intensity.

S.
21.3.2

21.3.2. *Neutron radiation*

The fission of the $^{235}_{92}U$ *isotope* is the result of the increased intensity of their neutron collapse.

In the natural circumstances of the *Uranium* element the proton electron and neutron process keep balance within the element. In the $^{235}_{92}U$ *isotope* the process is unbalanced: the event concentration of the *isotope* is less than the *Uranium* element ($Z_{U235} < Z_{U238}$) and consequently the intensity of the electron process of the *isotope* is more than the intensity of the *Uranium* element ($\varepsilon^e_{U236} > \varepsilon^e_{U238}$). As a result of the destruction of the *isotope*, the *fission-product-new-elements* remain without sufficient blue shift provision. They will seek any blue shift possibility available through the quantum membrane in the surrounding environment.

The neutron radiation is no other than the effect of the neutron collapse on the environment: the removal of blue shift from atoms of elements with blue shift surplus near to the fission process. There are no flying neutrons. The neutron radiation is the effect of the red shift of the neutron collapse of the atoms of the fission products with blue shift deficit.

S.
21.3.3

21.3.3. *Nuclear fusion*

The key is here the natural *plasma* status of the *Hydrogen*.

Ref
S.20.3

With reference to Section 20.3, the blue shift conflict of the accumulating *Hydrogen* atoms speeds up their motion up to $\lim i = c$. The collision at $\lim i = c$ destroys the *Hydrogen* atoms, which otherwise stable and withstand any blue shift impact from the *Quantum Membrane* and from other elements.

The distraction either results in
- neutron effect, since the free neutrons of the *Hydrogen* (with infinitely low intensity within the *Hydrogen*) will collapse and take massive blue shift from the *Quantum Membrane*; or

Ref
Tabl
20.1

- formulation of *Helium* with, with reference to Table 20.1, much higher electron process intensity.

Therefore the natural *plasma* contains two elements:
- *Hydrogen* as end product of the mass-energy transformation of the matter; and
- *Helium*, the formulating element, the element of the fusion, the result of the *Hydrogen* distraction.

The formulating *Helium* will be destroyed by the intensive blue shift impact from the *Quantum Membrane*.

The intensive blue shift impact of the *Quantum Membrane* is the consequence of the accumulation of the *Hydrogen* at the end of the quantum energy mass transformation process. The distraction will generate intensive blue shift and neutron effect. At the same time the *Quantum Membrane* guaranties the stability and the size of the *plasma*.

The *plasma* develops energy in the form of
- photons (generated by the free proton process); and
- radiated blue shift (generated by the *Hydrogen* and *Helium* electron processes and the distraction of the *Helium* element).

The *plasma* takes energy from the *Quantum Membrane* for the acceleration of the *Hydrogen* atoms and for the collapse of free neutrons.

22

The speed of light

Matter is the transformation of mass into energy and the energy into mass. The transformation establishes the time. Time means event and the event means time.

Photons of equal energy quantum compose the *Quantum System of Reference,* the acting *Quantum Membrane.*

Since Galileo we measure or try to measure the speed of light. This measurement in fact represents something more and something else than "just" the speed of light.

It is explained below by using H.L.Fizeau's experiment. Fizeau made the measurement in 1849 with rotating discs on a wheel with asymmetric holes at a certain distance from each other in overlap, as shown in Fig.22.1.

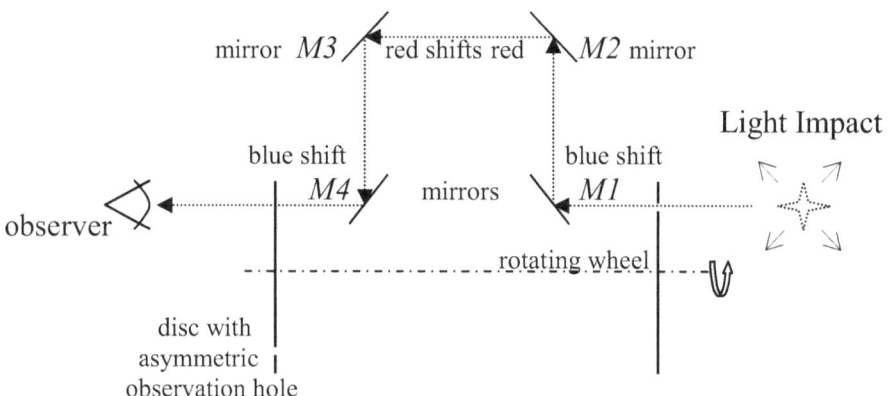

(The description of the measurement and the figure is taken from George Gamov and John M. Cleveland: Physics 1969.)

Fig
22.1

Fig.22.1

A light signal was sent towards rotating discs on a common wheel. Mirrors were used in order to lengthen the distance of the light "propagation" until detected on the other end of the rotating wheel.

The idea of the measurement was that the light would be detected depending on the speed of the rotating wheel. For detecting the light signal on the other end of the discs, the signal should make the distance between the asymmetric holes on the discs. The distance between the holes, the speed of the rotation and the length of the light path were known, so the speed of the propagation could be calculated.

The light signal was observed on the other side of the rotating discs. Should however the system of reference of the measurement (*Earth*) not have been in sphere symmetrical expanding acceleration, this light signal would have never been observed! Later experiments, like the Michelson-Morley one, have excluded the *motion* of the *Earth*, but not the accelerating [gravitational] one.

We take it that the *Earth* is at rest. Once the light signal has been generated, the *Quantum Membrane* is impacted. It is however a fiction, because no *Quantum System of Reference* exists with the *Earth* being at rest (meaning, without sphere symmetrical expanding acceleration). If the *Earth* is at rest, no signal can be detected.

Instead, the *Earth* is in sphere symmetrical expanding acceleration and the impact of the light suffers consecutive blue and red shifts on the mirrors. This is the only way the *Quantum Membrane* can convey the light impact. If the frequency of the light is f_o at the spot of the induction, its frequency at the first mirror will be, with reference to 15H4:

$$f_{M1} = f_o \left[1 + 1 - \sqrt{1 - \frac{(c-i)^2}{c^2}} \right] \qquad \text{22A1}$$

where $(c-i) = gt$ the acceleration of the system of reference.

At the second mirror the shift is red and it is red on the third as well. It gives on the third mirror a frequency of:

$$f_{M3} = f_o \left(1 - \sqrt{1 - \frac{(c-i)}{c^2}} \right)^2 \left[1 + 1 - \sqrt{1 - \frac{(c-i)^2}{c^2}} \right] \qquad \text{22A2}$$

Mirror 4 makes a blue shift impact and the detection is also a blue shift. It means the frequency of the light impact for observation is:

$$f_{obs} = f_o \left(1 - \sqrt{1 - \frac{(c-i)}{c^2}} \right)^2 \left[1 + 1 - \sqrt{1 - \frac{(c-i)^2}{c^2}} \right]^3 \qquad \text{22A3}$$

This frequency can only be detected through the hole of the second rotating disc, if the speed of the rotation allows the passing of the signal.

What we actually measure is the impact of the light signal, transferred by the *Quantum Membrane*. And the signal is only detectable, because of the motion of the *Earth*.

Photons are of *equal* energy quantum. There cannot be photons with more energy, passing others in "mission" carrying the impact of the light (or other) signals. This would mean that there were photons with energy more than the energy of the quantum entropy.

The measured $c = 299,792,458$ m/sec is *not* (just) the speed of light, rather it is the speed of the reaction (the transfer) of the *Quantum Membrane*, once impacted.

The "speed of the transfer" of the *Quantum Membrane* is obviously based on the speed of photons and is indeed c. Photons collide with each other and all of them equally have virtual mass less than the mass of the quantum entropy.

In our *interpretation* c is the speed of light. We call it "the speed of light", but in fact, all impacts, not just light, have the same effect, namely, the speed of the "reaction" of the *Quantum Membrane*. The *Quantum Membrane*, the sensitivity of a single quantum, conveys the impact with c, the "*speed of light*".

23

The energy benefit of the motion of the *Earth*

The sphere symmetrical expanding acceleration of the *Earth*, the motion with $i = \lim g\Delta t = c$ is balanced by the *Quantum Membrane*.

With reference to Sections 9 and 15.1, an impact on the *Quantum Membrane*, frequency of f – "hit" of n photons for a unit period of time at level h above the surface of the *Earth* – will be blue shifted by the acceleration of the *Earth*.

The impact, energy intensity of

$$e = \dot{m}c^2 = f \cdot q \qquad \text{23A1}$$

reflected from the surface of the *Earth,* will be

$$e_R = \dot{m}c^2\left(1 + 1 - \sqrt{1 - \frac{v^2}{c^2}}\right) = f \cdot q + f \cdot q\left(1 - \sqrt{1 - \frac{v^2}{c^2}}\right) \qquad \text{23A2}$$

- $\dot{m}$ is the measured mass equivalent of the impact within the system of reference of the *Earth*;
- $v = g \cdot t$ is the speed increase with acceleration g at distance h, the height of the impact above the *Earth*;
- t is the estimated time of this acceleration.

With reference to Sections 9 and 15.1, v, the relative speed and t, the time component are calculated from formula in 3C3: $h = \frac{c^2}{g}\left(1 - \sqrt{1 - \frac{v^2}{c^2}}\right)$

The key here is not the time during which the *Quantum Membrane* transfers the impact to the surface of the *Earth* by speed c. The key is the mass-energy and the quantum-energy difference of the spots of the impact and the reflection. The energy difference between these spots is equal to the work, what a mass equivalent of the impact would need in order to be accelerated by g on the distance between these spots.

The assessment is based on the constancy of absolute energies. We compare the quantum energy and mass (energy) values at the moments of the impact and the reflection.

The *Earth* is in sphere symmetrical expanding acceleration, in motion with $i = \lim g\Delta t = c$. This speed is constant. It cannot be more than i, but its *acceleration* – in order to keep this speed against the permanent impact of the *Quantum Membrane* – is also constant.

The *Quantum Membrane* transfers the impact but the photons are not accelerating. They are of equal energy quantum.

Our relativistic approach allows us to trust to our senses and observations. The sphere symmetrical expanding acceleration for infinite time, the motion with $i = \lim g\Delta t = c$ may be considered as stationary status. As a consequence, the quantum impact (as in the current case) – and any subject in "free fall" – can be considered as being in acceleration relative to the "stationary" *Earth*. This false perception prevents us from recognising and accepting the proven facts on the mass energy balance of the matter.

The acceleration of mass by external energy source needs external work. This work is proportional to the value of the acceleration and the mass. Since the work is external, the value of the mass stays constant during the acceleration. Subjects, with different mass values, driven by the same external energy source of the *gravitation* of the *Earth* could not have the same g value of acceleration.

The sphere symmetrical expanding acceleration of the *Earth* is a motion with constant $i = \lim g\Delta t = c$ speed. This is however an acceleration with g *relative* to the photons of the *Quantum System of Reference* and/or to any subjects in "free fall". As a result, the speed difference which the *Earth* gathers during its acceleration is real. Photons of the quantum impact and subjects with real mass in "free fall" are the recipients of the energy impact in collision with the *Earth*.

23A3   The impact in 23A1 in absolute terms would be: $mc^2 = f \cdot H$

where H is the Planck constant (on the *Earth*) $[\,Joule \cdot \sec\,]$

23A1 is equivalent to

23A4 $\dfrac{dE}{\varepsilon dt} = \dfrac{dmc^2}{\varepsilon dt} = \dfrac{dn}{\varepsilon dt}q\,;$ and $E = \dfrac{dn}{dt}q \cdot dt = f \cdot H\,;$ or $\dfrac{e}{\varepsilon} = \dfrac{\dot{m}c^2}{\varepsilon} = \dfrac{f}{\varepsilon}q$

The intensity and the time flow on the *Earth* is taken as: $\varepsilon = 1$ and $dt = 1$. With the substitutions, 23A1 turns into 23A3. All systems of reference have their own time flow and we measure intensities instead of absolute values in systems of reference.

With reference to 23A1, 23A2 and 15H6, the original energy intensity impact is increased by the *blue shift* of the *Earth*.

23A5 $e_R > e$ and $f_R > f$

The reason for the increase of the energy intensity and the frequency is the slow down of the time flow at the spot of the collision of photons with the surface of the *Earth*. With reference to 15D3 it is:

23B1 $dt_R = \dfrac{dt}{2 - \sqrt{1 - \dfrac{v^2}{c^2}}}\,;$ and $dt_R < dt$

The meaning of t and dt and dt_R are different.

t is a real time measurement and relates to 23A2. dt, dt_R both relate to the systems of reference of the *Earth* but at different spots of the impact and collision with the *Quantum Membrane*. They characterise the energy status together with h, the height of the impact or together with v the value of the speed difference of the acceleration with g for time t.

How can it be that with the motion of the *Earth* $i = \lim g\Delta t = c$ constant, the time flow is different than $dt = 1$? We can characterise the sphere symmetrical expanding acceleration of the *Earth* by its intensity. The intensity in fact is equal to the actual value of the acceleration.

For the case without collision, it is: $\dfrac{c-i}{dt} = g = a = \varepsilon$ 23B2

At the spot of the reflection the local intensity (acceleration) of the motion, as a consequence of the collision, is higher. It must be, otherwise the speed of the motion would be less than $i = \lim g\Delta t = c$. The effect of the impact is compensated by the increased intensity (of the acceleration):

$$\varepsilon_R > \varepsilon \quad \text{since} \quad \frac{c-i}{dt_R} = \frac{c-i}{dt}\left(2 - \sqrt{1 - \frac{v^2}{c^2}}\right) = a_R = \varepsilon_R ;$$ 23B3

$v \neq (c - i)$ - they represent different speed values: v relates to the energy difference of the spots; $(c - i)$ is the constant speed difference of the acceleration of the *Earth*. We need it to characterise the intensity of the acceleration, the motion with $i = \lim g\Delta t = c$.

23B2 and 23B3 mean that the product of the intensity and the time shall give a constant value of $(c - i)$. Here, the dimension of the measurement is of secondary importance. The important thing is that the intensity of the event is proportional to the value of the acceleration. The change in the value of the acceleration is a consequence of the collision of the *Quantum Membrane* with the surface of the *Earth*.

Addressing our concern above about the constancy of the speed of the motion and the changing time flow, we have to note that the *constant* speed of the sphere symmetrical expanding acceleration is the result of *permanent* acceleration. In the case of $i = \lim a\Delta t = c = const$ the time flow relates to the actual acceleration. The value of the acceleration is in fact the intensity of the event, and this determines the time flow.

The n number of photons (of equal energy quantum) at the impact at h and at the reflection on the surface of the *Earth* is obviously the same. The absolute energies, as they should be, are equal in both cases. As proof, with reference to 23A1, 23A2 and 23B2, 23B3:

$$E = E_R; \quad \text{and consequently:} \quad \frac{e}{\varepsilon} = \frac{e_R}{\varepsilon_R}; \quad \text{or} \quad \frac{\dot{m}c^2}{\varepsilon} = \frac{\dot{m}_R c^2}{\varepsilon_R};$$ 23B4

The effects of the impacts, however, are different. With reference to 15E5, the energy intensity of the impact, reflected from the surface of the *Earth* and measured (detected) at the level of the original "hit" is:

$$e_D = \dot{m}c^2 \sqrt{1 - \frac{v^2}{c^2}}\left(2 - \sqrt{1 - \frac{v^2}{c^2}}\right) = f \cdot q \sqrt{1 - \frac{v^2}{c^2}}\left(2 - \sqrt{1 - \frac{v^2}{c^2}}\right)$$ 23C1

The value of v is the same as in 23A2, since the distance is the same.

With reference to 15H7, the blue shifted energy intensity and frequency at the detection are less than the original:

23C2
$$e_D < e \quad \text{and} \quad f_D < f$$

With reference to 15F2, the time flow at the detection is "quicker" than the time flow of the *Earth*:

23C3
$$dt_D = \frac{dt}{\sqrt{1 - \dfrac{v^2}{c^2}\left(2 - \sqrt{1 - \dfrac{v^2}{c^2}}\right)}}; \qquad dt_D > dt$$

23C4
$$\frac{c-i}{dt_D} = a_D = \varepsilon_D = \frac{c-i}{dt}\sqrt{1 - \frac{v^2}{c^2}\left(2 - \sqrt{1 - \frac{v^2}{c^2}}\right)}; \quad \text{and} \quad \varepsilon_D < \varepsilon$$

The detector is faced towards the *Earth*. The red shift will decrease the acceleration of the spot toward the *Earth*. The speeded up time flow at the spot of the detection is the reason for the lower value of the energy intensity and the frequency.

With reference to the assumptions on the relation of the acceleration, intensity and time flow and the constancy of absolute energies, with reference to 23B4:

23C5
$$E = E_R = E_D; \qquad \text{and} \qquad \frac{\dot{m}c^2}{\varepsilon} = \frac{\dot{m}_R c^2}{\varepsilon_R} = \frac{\dot{m}_D c^2}{\varepsilon_D}$$

With reference to 15I3, the difference between the detected (*red* shifted) and the reflected (*blue* shifted) impacts – the *energy intensity benefit* of the impact – at the level of the original impact is:

23D1
$$\dot{m}_D c^2\left(1 - \sqrt{1 - \frac{v^2}{c^2}}\right) = -\dot{m}c^2\left(1 - \sqrt{1 - \frac{v^2}{c^2}}\right)\left(2 - \sqrt{1 - \frac{v^2}{c^2}}\right)$$

The minus sign demonstrates that this is a *receipt* of energy intensity by the detector.

Resolving 23D1, the mass equivalent of the detected energy intensity in 23A1 is:

23D2
$$\dot{m}_D = \dot{m}\left(2 - \sqrt{1 - \frac{v^2}{c^2}}\right)$$

The surplus of the reflected *(blue shifted by the Earth)* and detected *(red shifted at the detector)* energy intensity at the level of the original impact – in measured mass equivalent (effect) – is:

23D3
$$\Delta\dot{m} = \dot{m}_D - \dot{m} = \dot{m}\left(1 - \sqrt{1 - \frac{v^2}{c^2}}\right)$$

23D3 means that the $e = \dot{m}c^2$ intensity impact in 23A1 on the *Quantum Membrane* after *blue shifted* by the *Earth* and *red shifted* by the detector results in $\Delta\dot{m}$ mass equivalent intensity benefit:

$$\Delta\dot{m} = \dot{m}\left(1 - \sqrt{1 - \frac{v^2}{c^2}}\right); \qquad\qquad \text{23D4}$$

The energy intensity benefit at the detector is the intensity of the work which accelerates mass $\dot{m}$ on speed v with acceleration g:

$$e_\Delta = \Delta\dot{m}c^2 = \dot{m}c^2\left(1 - \sqrt{1 - \frac{v^2}{c^2}}\right); \qquad\qquad \text{23D5}$$

With reference to 15G1, 23B4 and 23C4, the absolute energy values of the process are perfectly *equal*. The benefit is *energy intensity*, the measured effect of the event, function of the motion and the time flow.

With reference to 15J2, there could be a different height level to be found, below the original impact, where the result of the detected and red shifted frequency is equal to the original one.

Similarly to the event of *fire* or the *hydrocarbon* and *nuclear* engines, the benefit is *energy intensity*, the measured effect of the events. Energy and work intensities, breaking the mass-energy balance of atoms and molecules are tiny values in their own elementary systems of reference. Their effect is rather significant in a system of reference with slower time flow.

In this *energy-intensity-benefit* case, the point is not about the *value* of the resulted energy intensity. It is about the fact of gaining – in our conventional understanding – *energy* for use.

The *Earth*, in sphere symmetrical expanding acceleration, is in genuine blue shift with the *Quantum Membrane*. The blue shift is the result of balance, the effect of the acceleration at constant ($i = \lim g\Delta t = c$) speed for infinite time.

An impact to the *Quantum Membrane*, different than the sphere symmetrical expanding acceleration of the *Earth*, is an additional "artificial" impact. It works against the sphere symmetrical expanding acceleration of the *Earth*. At the spot of the impact, the *Earth* compensates and the motion is kept $i = \lim g\Delta t = c = const$: a more intensive acceleration, a slow down of the time flow. The direct consequence is an increase of frequency – additional blue shift.

The reflected impact from the surface of the *Earth* with frequency, higher than the original, is transferred by the *Quantum Membrane* and collides with the detector, faced towards the surface of the *Earth*.

If the detector is in motion together with the *Earth*, the detected impact will be red shifted. The energy intensity and the frequency of the impact at the detector, after the red shift, is less than those of the original impact. The red shift at the detector slows down the acceleration toward the surface of the *Earth*, decreases the energy intensity and speeds up the time flow. The direct consequence is the decrease of the frequency.

The motion of the *Earth* impacts the *Quantum Membrane* and creates $e_\Delta = \Delta \dot{m} c^2$ *quantum energy*. This source of energy is there for us to use.

Attachment

Executive Summary
of
The Energy Balance of Relativity

Relative motion considers at least two systems of reference: the supposed to be stationary one and the one in motion. Events that may happen in the system of reference in motion certainly happen in the stationary system as well. The general rule of any relativistic consideration is that events must be examined from the point of view of all systems of reference, where they simultaneously occur. And the results can only be accepted as correct if they comply with the laws of physics from the point of view of *all* systems being examined.

We investigate in this book the energy balance of systems of reference in relative motion with the final objective of comparing a system of reference in acceleration for infinite time with *Gravitation*.

The investigation is conducted in three stages:

Chapter I is the review of the Special Theory and the review of the foundation of the General Theory of Relativity.

Chapter II is the introduction of the thesis. This chapter gives the formula for the time relations of systems of reference in relative motion, characterises the unity of the mass-energy balance, defines categories of intensity of events and event concentration, describes the motion with $\lim v = c$, the acceleration for infinite time, investigates the blue and red shift of electromagnetic waves and gives the premium formula of the blue and red shift sequence for use.

Chapter III is an outline of a hypothesis.

Chapter I

The *general problem* with the Special Theory of Relativity is that the real *reciprocal character* is missing from the concept. Two systems of reference in relative motion are equal parts of the same relation. The time relations must be valid from the point of view of both systems of reference, independently of which one is taken as the supposed to be stationary system of reference and which one is considered as being in motion. Because of the missing reciprocal character, the time formula in the Special Theory for systems of reference in relative motion is inadequate.

There is also a *misunderstanding* and misinterpretation of the transformation of space coordinates in the Special Theory. Distances, *measured* in systems of reference in relative motion, are different, but not because inert bodies or systems of reference in motion undergo contraction or change of any kind in their geometrical size. *De facto* static geometrical data are constant and invariant. The coordinates of these data are measurements, e.g. results of events, function of the time flow, depending on the relative, supposed to be status of the systems of reference. Therefore, the measured coordinates are variant.

The energy balance proves that the collision of electromagnetic waves with inert bodies or systems of references in acceleration in a space without gravitational field results in a similar effect as Einstein *a priori* attributed to Gravitation in his paper "On the influence of Gravitation on the Propagation of Light" of 1911. The energy balance also shows the natural correction of the frequency of electromagnetic waves in collision. Any adjustment of the speed of light in this case is not just unnecessary, but obviously makes no sense.

The consequence of the collision of electromagnetic waves with inert bodies or systems of reference in motion is an energy exchange. The collision from in front results in a certain increase of the frequency of electromagnetic waves, while the inert body or system of reference in motion gives off part of its kinetic energy in the same value. The collision from behind means the opposite: electromagnetic waves give off part of their radiation energy and the kinetic energy of the inert body or system of reference in motion increases by the same value.

The proper description of the energy exchange of the collision extends the meaning of Doppler's formula and opens further dimensions for its usage. It characterises all phases of the energy-mass transfer including expanding acceleration and accelerating collapse.

The incorrect time formula in the Special Theory, the misunderstanding of the relativistic effect on the space coordinates and, in parallel, the *a priori* conclusion on the effect of Gravitation lad Einstein to the statement in his paper on "The Foundation of the General Theory of Relativity" in 1916, that the Euclidean geometry can not be applied to describe rotating space-time continuum systems of reference.

The chapter is going to prove that while space coordinates can not be projected in a conventional way, indeed, there is no problem with the Euclidean geometry. The approach is what must be changed. The proper time formula itself resolves the issue. The adequate relativistic use of the space coordinates, the consequence of the time relations, shows the Euclidean geometry holds good and results in a proper description even for non-uniform systems of reference in rotation, the expanding and rotating space-time continuum.

While no event (motion) means: no time definition, the motion modifies the time flow, expands the space and the key parameter is the *acceleration,* the valid alternative for understanding Gravitation.

The main objective of this book is the thesis in Chapter II, but before starting with this, an important question must be addressed: how can a hundred-years-old theory, proven by experiments, be questioned? Our concerns relate to the explanations, the interpretation of the experimental results.

Chapter II

This chapter introduces the thesis.

The establishment of the correct time relation between systems of reference in relative motion is crucial. Events happen in time and the use of an improper time formula results in an incorrect energy balance.

The motion modifies the time flow. Time flows faster within systems of reference in motion. The proper time formula is the inverse of what is suggested by the Special Theory.

$$d\tau_{(motion)} = \frac{dt_o}{\sqrt{1 - \frac{v^2}{c^2}}}$$

The first step on the road to revise the principle of equivalence is to answer the question: does the acceleration, the motion with *a=const,* modify the time flow?

The answer is: Yes, it does. The acceleration speeds up the time flow. The differential form of the equation predicts a non-linear time flow and path description:

$$\frac{d\tau_a}{dt_o} = \frac{1}{\sqrt{1 - \frac{a^2 t_o^2}{c^2}}}$$

$$\tau_a = \frac{c}{a}\arcsin\frac{at_o}{c}; \qquad S_a = \frac{c^2}{a}\left(1 - \sqrt{1 - \frac{a^2 t_o^2}{c^2}}\right)$$

The time flow in systems of reference in motion is modified, because three dimensional systems of reference, while moving in space, do not and can not influence the propagation of light. The *description* of the propagation of light, therefore in the system of reference in motion is longer than in the stationary one. The bigger product of $c \cdot \Delta t$ with *c=const* results in a longer duration, consequently, a faster time flow. The description of the light propagation (its observation) is declined in the given system of reference in motion and gives a longer path than in the stationary system of reference.

Light has no vector components. Any supposed light vector component is equal to the genuine value of the light vector. Therefore, its velocity is one and the same in any direction. It means the time relation in a given system of reference in motion is independent of the direction of the motion.

What is the work necessary to accelerate a system of reference? There is another question, no less important to be answered: why is it necessary to investigate this when it can be found in every textbook?

Because we are looking for the work-formula which corresponds to the laws of physics from the point of view of both systems of reference, the one in acceleration and the other one in which the acceleration takes place. While the value of the acceleration from the point of view of the stationary system of reference is constant, it is variant from the point of view of the system of reference in acceleration.

To establish the correct work formula of the acceleration for both systems of reference, the acting force value, product of $m \cdot a$ must be adjusted. And it leads to the definition of a *relativistic* or *acting* mass.

$$dm_o = \frac{1}{\sqrt{1-\dfrac{v^2}{c^2}}} dm_v$$

$$dm_o = \frac{1}{\sqrt{1-\dfrac{a^2 t_o^2}{c^2}}} dm_a$$

The value of the relativistic mass at $v=0$, at rest, is equal to the inert mass: $m_{inert} = m_{rel}$

$$W_a = m_a c^2 \left(1 - \sqrt{1-\frac{a^2 t_o^2}{c^2}}\right)$$

$$\Delta t_o = t_o - 0$$

The definition of the *relativistic mass* makes it possible to determine the absolute value of work necessary for accelerating a system of reference with $a = const$. As expected, it is equal from the point of view of both systems of reference.

This work formula using the inert mass also gives the Newtonian equation. It proves there is only one law of *physics* to apply. The origin is the same, the expressions are different.

$$W = m\frac{a^2 t_o^2}{2}$$

Einstein's work and energy formulas are in fact *intensities,* their actual appearance in the given system of reference. The work intensity expression for the stationary system of reference gives it in its well-known shape.

$$w_a = \frac{m_a c^2}{\sqrt{1-\dfrac{a^2 \Delta t_o^2}{c}}} - m_a c^2$$

The considerations about inert and relativistic masses bring up important findings:

$E_{rest} = mc^2$ is the value of the full energy of the inert mass at rest without any motion;

$E_{motion} = mc^2$ is also the full value of the total kinetic energy. Since the inert mass in full motion *(when v=c)* has no meaning, it is presented by its *virtual* value.

The *energy-mass unity* is in dynamic balance between these two ends. The *energy reserve at rest* is a new category for characterising the status between these end points.

$$m_v c^2 \sqrt{1 - \frac{v^2}{c^2}} = m_v c^2 - W$$

The inert mass value characterises the total *absolute* energy at rest. The relativistic mass value characterises the *actual* energy of the mass in motion. In everyday practice we measure relativistic masses.

The motion modifies the time flow and this raises a new question to be answered: does the modified time flow change the energy demand of events?

The one and the same event must have the same absolute energy in any system of reference in motion.

$$\frac{dW}{dt_o} = w_o \quad \text{and} \quad \frac{dE}{dt_o} = e_o$$

But *systems of reference*, distinct in motion, are different in their time flow, therefore, they must have *different intensities* in their energy usage. Events do not just happen "faster or slower" in systems of reference distinct in motion, but with "lesser or higher" intensity. Otherwise the energy balance could not be maintained. Consequently, their *event concentration* is different.

$$\frac{dW}{d\tau_v} = w_v \quad \text{and} \quad \frac{dE}{d\tau_v} = e_v$$

$$\frac{dW}{d\tau_a} = w_a \quad \text{and} \quad \frac{dE}{d\tau_a} = e_a$$

The mathematical shape of *work* and *energy intensities* is similar to that of capacity. There is, however, a principal difference between the two. The capacity only relates to a certain event and shows the value of work spent for a particular period. The *intensity* of events for a system of reference is uniform and relates to all events within the given system of reference.

The *intensities* are the real appearances of events within the systems of reference. Together with the *event concentration* they relate exclusively to and characterise the systems of reference.

The *event concentration* helps us to describe the *muon escape,* one of the practical manifestations of the existing relativity, giving the correct explanation for finding cosmic muons on the surface of the *Earth.*

The principle of equivalence is re-examined. The acceleration is taken for infinite time.

This is a motion with speed $i = \lim a \Delta t_o = c$, with constant energy and permanent work demand, with the speed of the motion dropping down, fluctuating constantly.

$$e_{oi} = \frac{m_i c^2}{\sqrt{1 - \dfrac{i^2}{c^2}}} = const$$

In order to keep the speed and the energy of the motion quasi constant and compensate for the drop-down, work must be envisaged.

$$W_{a(drop-down)} = m_i c^2 \left(1 - \sqrt{1 - \frac{a_{(n)}^2 \Delta t_o^2}{c^2}} \right)$$

Therefore, this motion is also a permanent acceleration.

Thus, the motion with $i = \lim a \Delta t_o = c$, the acceleration for infinite time is unique and exclusive. It needs permanent work, and, thus, creates working capability and, in spite of the work spent, its energy remains constant.

System of reference in motion with $i = \lim a \Delta t_o = c$, the acceleration for infinite time, can be used as a *basic platform* to compare systems of reference in motion with each other. The comparison through a basic platform is an important tool to find the correct relation and to exclude the paradox of relativity.

The motion with $\lim i = c$ indicates a question: what is the maximum summarised equivalent speed of an inert body or system of reference in motion with *v=const,* achievable within a system of reference in motion with $i = \lim a \Delta t_o = c$? It is always less than *c,* the speed of light.

$$u = c \sqrt{1 - \frac{(c^2 - i^2)(c^2 - v^2)}{c^4}}$$

After all this background we arrive at the examination of the *Pound-Rebka-Snider experiment,* the measured *blue shift* of photons, caused by the effect of the supposed to be *Gravitation.* And the question is posed: may a system of reference in motion with $i = \lim a \Delta t_o = c$, the acceleration for infinite time, result in *blue shift* in collision with light photons?

And the answer is yes! The collision of electromagnetic waves with a system of reference in motion with $i = \lim a \Delta t_o = c$, the acceleration for infinite time, results in *blue shift* of light photons.

$$\Delta E_\gamma = E_\gamma \left(1 - \sqrt{1 - \frac{g^2 \Delta t^2}{c^2}} \right)$$

The calculated value of the *shift* within a system of reference *with no* gravitational field, but in motion with $i = \lim a \Delta t_o = c$, co-relates well with the measured results.

$$\frac{\Delta E}{E_\gamma} = 2.442 \cdot 10^{-15}$$

Thus, a system of reference *without any gravitational field* may cause *blue shift* and, of course also *red shift*. The comparison with a shift, which is supposed to be caused by *Gravitation* is unavoidable. And there is a point to note: photons do not accelerate and can not be accelerated. Why?

Because the energy of full motion and full rest of photons are equal, just their energy-mass status is different. The energy of motion means full energy without any mass, the energy at full rest means the mass itself.

$$E_{motion} = mc^2 =$$
$$= E_{rest} = mc^2$$

There is no room for the increase of the energy of the photon from its own sources. The transformation of the inert mass into energy with reaching the speed of light is complete.

The *source* of the energy surplus and energy deficit of the *blue* and *red shifts* is the kinetic energy of the system of reference in motion with $i = \lim a \Delta t_o = c$. It gives off part of its energy in the case of the *blue shift* and takes off part of the energy from the radiation in the case of the *red shift*. And we need not to be surprised, since this recognition is in full compliance with what Einstein stated in his ground-breaking paper in 1905 where he asked 'Does the Inertia of a Body depend upon its Energy-content?' Yes, it does. "If the theory corresponds to the facts, radiation conveys inertia between the emitting and absorbing bodies." (Quotation from Einstein, Source: "Does the Inertia of a Body depend upon its Energy-content?" Principle of Relativity, Dover Publications, Inc).

100 years' experience proves this statement perfectly corresponds to the facts.

The expression of the value of the blue shift through the virtual value of the photon's inert mass gives the gravitational potential.

$$\Delta E_\gamma = E_\gamma \left(1 - \sqrt{1 - \frac{g^2 \Delta t^2}{c^2}} \right) =$$
$$= mgh$$

In other words, the *blue* and *red* shifts are linearly proportional with the altitude in a space without gravitational field, as is the gravitational potential.

Here we have found the reason for the *drop-down* of the energy of the motion with $i = \lim a\Delta t_o = c$, that necessitates the work to be envisaged. It is for compensating for the energy taken off by the blue shift of photons.

The sequence of *blue and red* shifts of electromagnetic waves results in energy transfer, the transfer of the *energy surplus* provided by the kinetic energy of the system of reference in motion with $i = \lim a\Delta t_o = c$, the acceleration for infinite time. The surplus premium, taken off by the *blue-red shift* sequence to be utilised, is:

$$\Delta E_{(red-blue)} = \Delta E_{(red)} - \Delta E_{(blue)} = E_\gamma \frac{g^2\Delta t^2}{c^2} \cdot \frac{1-\sqrt{1-\left(g^2\Delta t^2/c^2\right)}}{1+\sqrt{1-\left(g^2\Delta t^2/c^2\right)}} > 0$$

The above equation is the main message of this thesis.

The motion with $i = \lim a\Delta t_o = c$ raises the question about the value of the frequency of the collision, predicted by Einstein to be of an infinite value. Doppler's formula defines it as a frequency, quasi equal to the frequency of light, corresponding to the speed difference between c and $\lim i = c$.

$$f_E = f\sqrt{\frac{1+\dfrac{|c-i|}{c}}{1-\dfrac{|c-i|}{c}}} \cong f$$

Chapter III

This chapter is a hypothetical outline. It has three sections with different hypothetical messages.

Section 10 summarises the arguments and findings and states: *the sphere symmetrical expanding acceleration of the Earth, the motion with* $i = \lim a\Delta t_o = c$ *for infinite time,* is a realistic alternative for the understanding of the meaning of *Gravitation.*

-   *Gravitation* is *energy*, the supposed to be system of reference, which provides the energy for the *sphere symmetrical expanding acceleration of the Earth*, the system of reference *in motion with* $i = \lim a\Delta t_o = c$ for infinite time.

-   The *sphere symmetrical expanding acceleration of the Earth, a motion with* $\lim i = c$ obviously *modifies* our existing view on the *gravitational free-fall.*

- The collision of electromagnetic waves with the accelerating surface of the *Earth*, the blue shift, *increases* the energy of the electromagnetic waves, but *most importantly*, the kinetic energy of the *Earth* in sphere symmetrical expanding acceleration, motion with $i = \lim a\Delta t_o = c$ can be transformed through *blue and red* shift sequence *into energy to be used.*

With Section 10 the thesis introduced in Chapter II can be considered as completed.

Section 11 is an outline demonstration of the energy balance of the hypothetical sphere symmetrical expanding acceleration of the *Earth* and light photons. *Gravitation* provides the energy and light photons control the process and keep the balance. *Earth is in sphere symmetrical expanding acceleration around and away from its centre.*

In order to quantify the dimension of *i* the speed of the sphere symmetrical acceleration and *z*, the *event concentration* on the surface of the *Earth* an assessment of first approximation was performed and values found:

$$i = c\sqrt{\frac{6\cdot10^{24}}{6\cdot10^{24}+2.96\cdot10^{-6}}} \quad \text{which is indeed } \lim i = c$$

$$z = \sqrt{1 - \frac{6\cdot10^{24}}{6\cdot10^{24}+2.96\cdot10^{-6}}} \quad \text{which is indeed } \lim z = 0 \text{, practically zero}$$

Section 12 gives the outline of a hypothesis of the *sphere symmetrical expanding acceleration* and *accelerating collapse* in general, a *pulsation of the energy-mass unity* for infinity.

The description of the pulsation is given in accordance with the adjusted (comprehensive) Doppler formula introduced in Chapter I. It establishes the energy-mass status of the *Black Hole,* the pure energy concentration without mass. The other end of the process is also introduced, identifying it as *White Room,* the fully exploited energy, the fully expanded status of the mass at rest.

I am grateful to my wife for her permanent support and to
Keith (William) Hardwick for his encouragement and for reviewing and
correcting my English.

www.ingramcontent.com/pod-product-compliance
Lightning Source LLC
Chambersburg PA
CBHW081125170526
45165CB00008B/2557